# SpringerBriefs in Computer Science

More information about this series at http://www.springer.com/series/10028

Maurizio Martellini · Stanislav Abaimov
Sandro Gaycken · Clay Wilson

# Information Security
# of Highly Critical Wireless
# Networks

 Springer

Maurizio Martellini
Landau Network Fondazione Volta
Milan
Italy

Stanislav Abaimov
University of Rome Tor Vergata
Rome
Italy

Sandro Gaycken
Digital Society Institute
European School of Management and
   Technology (ESMT)
Berlin
Germany

Clay Wilson
Cybersecurity Studies Graduate Program
University of Maryland University College
Largo, MD
USA

ISSN 2191-5768          ISSN 2191-5776   (electronic)
SpringerBriefs in Computer Science
ISBN 978-3-319-52904-2          ISBN 978-3-319-52905-9   (eBook)
DOI 10.1007/978-3-319-52905-9

Library of Congress Control Number: 2017930284

Printed on acid-free paper

This Springer imprint is published by Springer Nature
The registered company is Springer International Publishing AG
The registered company address is: Gewerbestrasse 11, 6330 Cham, Switzerland

# Contents

# Chapter 1
# Introduction

Three industrial revolutions were catalyzed by technology advances of the last three hundred years of human evolution. With the breakthrough in computer engineering and industrial automation, the beginning of the XXI century is witnessing such phenomena as Internet of Things, Robotics, Virtual Reality, Cyber Warfare, and Industry 4.0. The emerging technologies are initiating the fourth wave of technological breakthrough, the so called Fourth Industrial Revolution.

Global smart architectural interconnectivity, the current reality of the human world, comprises smart machines in home, office, production and military facilities, earth and space critical infrastructure. Industry 4.0, perceived as automation and data exchange in manufacturing, communication, and control technologies, includes cyber-physical systems able to wirelessly monitor and control processes through smart sensors. The wireless technologies are easy to use in communication and data transfer, and Highly Critical Wireless Networks are now part of every military, industry, and office environment.

Designed to speed up the work efficiency, interdependencies and complexities of industrial and corporate wireless systems, generate multiple cyber security vulnerabilities. The access to wireless devices provides immediate penetration to internal networks, and in highly critical networks even the lowest unauthorized privileged access can compromise the mission. The increasing sophistication, accidental or intentional misconfigurations of equipment, and exponentially growing number of vulnerabilities urge for comprehensive research, monitoring, assessment, and testing of the wireless equipment and software.

One of the most efficient protection measures is proactive cyber security testing, which detects and classifies flaws in cybersecurity. With a more sophisticated design of wireless technologies, their security testing is more complicated and includes additional measures, such as acknowledgement of detectability and vulnerability of routers and adapters to further develop and deploy preventive measures against "eavesdropping," denial of service, security breaches, and unauthorized remote control of wireless devices.

© The Author(s) 2017
M. Martellini et al., *Information Security of Highly Critical Wireless Networks*,
SpringerBriefs in Computer Science, DOI 10.1007/978-3-319-52905-9_1

Intelligence attackers, cybercriminals, and cyber terrorists, with the level of preparation equal to the level of technologies, range from trained military experts with access to supercomputer technology to teenagers with smartphones downloading hacking applications.

In cybersecurity, defending is always more difficult than attacking, as the defenders have to secure every single vulnerability, while attackers need only one to breach the defenses. And there is no certain way to discover every vulnerability in the system and network.

This brief will introduce the reader to the vital elements of Highly Critical Wireless Networks, relevant international and national regulations standards, latest cybersecurity events, modern security solutions, and possible future cybersecurity challenges.

The main idea underpinning this brief is that, up to now, there is not a single-bullet solution to enhance the security and resilience of the Highly Critical Wireless Network seen, by raising a medical analogy, as the "central nervous system" of a forthcoming fully digitalized world of "human being and things." Besides the obvious problems, among others, related to the freedom of the web and the absence of a universal convention dealing with the governance of the Highly Critical Wireless Networks, there exists the difficulty to develop cost-effective security scenarios dealing with all the possible vulnerabilities of the wireless networks.

The Goals and Objectives of the brief are set to review the current and future cybersecurity challenges in wireless technologies, and their cybersecurity testing practices.

The target audience of the paper is cybersecurity testers, cyber security auditors, cybersecurity and network architects, security managers, software developers, and systems and network administrators.

# Chapter 2
# What Is Highly Critical Wireless Networking (HCWN)

A Highly Critical Wireless Network (HCWN) may consist of several interconnected communications devices used to support a critical function. HCWN are commercial wireless systems that use the TCPIP digital protocols, which are beginning to dominate our communications infrastructure and reshape our industrial base. Variations on these TCPIP communications protocols, using high or low power signals, have names such as Wi-Fi, Bluetooth, ZigBee, Z-Wave, cellular mobile radio, or mesh networks. Highly Critical Wireless Networks are now part of every industry, and are fast becoming required features in our new cars, airplanes, utilities, and work environments. HCWN technology can also be found in widespread use throughout developing countries in Africa and Asia where construction of a wired infrastructure was skipped in favor of deployment of mobile equipment and other digital devices that connect wirelessly to the Internet [1].

## 2.1 ZigBee

ZigBee is a wireless communications technology that is relatively low in power usage, data rate, complexity, and cost of deployment. It is an ideal technology for smart lightning, energy monitoring, home automation, and automatic meter reading, etc. ZigBee has 16 channels in the 2.4 GHz band, each with 5 MHz of bandwidth [2]. ZigBee is considered as a good option for metering and energy management and ideal for smart grid implementations along with its simplicity, mobility, robustness, low-bandwidth requirements, low cost of deployment, its operation within an unlicensed spectrum, and easy network implementation.

© The Author(s) 2017
M. Martellini et al., *Information Security of Highly Critical Wireless Networks*,
SpringerBriefs in Computer Science, DOI 10.1007/978-3-319-52905-9_2

## 2.2  Z-Wave

Z-Wave is another wireless communications technology, considered as an alternative to ZigBee. Z-Wave was developed by the Z-Wave Alliance, an international consortium of manufacturers. The simple, modular, and low-cost features make Z-Wave one of the leading wireless technologies in home automation. Z-Wave can be easily embedded to consumer electronic appliances, such as lighting, remote control, and other systems that require low-bandwidth data operations.

## 2.3  Cellular Network Communication

Existing cellular networks can also be a good option for communicating between smart meters and the utility and between far nodes. The existing communications infrastructure avoids utilities from spending operational costs and additional time for building a dedicated communications infrastructure. Cellular network solutions also enable smart metering deployments spreading to a wide area environment.

## 2.4  Wireless Mesh Networks

A mesh network is a flexible network consisting of a group of nodes, where new nodes can join the group and each node can act as an independent router. A mesh network will add flexibility and efficiency to the Internet-of-Things, where all household devices will be connected for two-way communication to power suppliers. Mesh networks allow power meters or connected devices to act as signal repeaters. Adding more repeaters to the network can extend the coverage and capacity of the network. Mesh networks rely on HCWN, and can be used for complex metering infrastructures and home energy management. For example, T-Mobile's Global System for Mobile Communications (GSM) network is chosen for the deployment of Echelon's Networked Energy Services (NES) system. An embedded T-Mobile SIM within a cellular radio module will be integrated into Echelon's smart meters to enable the communication between the smart meters and central utility. Mesh networking systems are highly complex, and are self-organization, self-healing, self-configuring, and offer high scalability. A mesh network has a self-healing characteristic which enables the communication signals to find an alternate transmission route if any node drops out of the network. Each meter or device on a mesh network acts as a signal repeater until the collected data reaches the electric network access point. Then, collected data is transferred back to the electric utility through the power lines, or via the HCWN communication network.

Mesh networking improves network performance, balances the transmission load on the network, and extends the network coverage range. These features make it suitable for supporting the Internet of Things and the Smart Grid [2].

# References

1. Austad, W., & Devasirvatham, D. (2014, May 06). *How Wireless Networks Impact Security*. Retrieved from RadioSource International: http://www.rrmediagroup.com/Features/ FeaturesDetails/FID/450
2. Gungor, V. C. (2011). *Smart Grid Technologies: Communication Technologies and Standards*. Retrieved from http://home.agh.edu.pl/~afirlit/LabRSMSM/Wykl045%20Smart%20Grids% 20-%20communication%20Technologies.pdf

# Chapter 3
# Applications for HCWN

HCWNs are regularly used by first responders and the military to support communications in remote areas; they may also be used to control portable medical devices; they may connect remote locations with SCADA systems to power generators and other centrally located critical civilian infrastructures and facilities; or, they may be the communications system that controls and monitors household devices as they are part of the new Smart Electric Grid for power distribution.

## 3.1 Terrestrial Trunked Radio

TETRA is formerly known as Trans-European Trunked Radio, and is an example of a networked system for mobile HCWN operation. TETRA was designed for use by emergency services in remote locations, for rail transport staff train radios, and also for use by the military. TETRA uses Time Division Multiple Access (TDMA) with both single point and multi-point transmission to enable multiple simultaneous TCPIP sessions for digital data transmission. TETRA is used for mission critical networks, where all aspects of the communications links are designed to be redundant and fail-safe. Mobiles and portable devices can use TETRA in "direct mode" for walkie-talkies, and for rapid deployment (transportable) networking for disaster relief. Digital data applications for messaging, voice, and video can provide situational awareness to decision makers and emergency responders to help manage disasters or crisis scenarios. One advantage is that TETRA networks can be quickly provisioned to provide the necessary connectivity to support these applications [1]. At the end of 2009, over 114 countries reportedly were using TETRA systems in Western Europe, Eastern Europe, Middle East, Africa, Asia Pacific, Caribbean, and Latin America [2].

© The Author(s) 2017
M. Martellini et al., *Information Security of Highly Critical Wireless Networks*,
SpringerBriefs in Computer Science, DOI 10.1007/978-3-319-52905-9_3

## 3.2   Medical Devices

Technical advances have transformed the delivery of health care, and improved capabilities for better patient care through use of mobile devices that are connected through the internet. A medical device can now take the form of a mobile instrument, portable apparatus, portable implant, or other similar article used for remote medical monitoring. These portable devices are intended for use in the diagnosis or treatment of disease or other conditions.

This increase in use of mobile devices for monitoring has also resulted in more communications interconnectivity between remote mobile medical devices and other centrally managed clinical systems. More interconnectivity leaves portable medical devices open to the same types of vulnerabilities described here for other networked computers and digital communications systems. There is an increasing concern that the interconnectivity of these medical devices creates vulnerabilities that can directly affect clinical care and patient safety. Recently, the SANS Institute reported that 94% of health care organizations have been the victim of a cyberattack [3]. This includes attacks on mobile medical devices as well as against the interconnected central management critical infrastructure.

The integration of medical devices, networking, software, and operating systems means that medical devices are challenged by increased complexity, which opens more cybersecurity vulnerabilities [3]. Despite the potentially lethal impact of compromised medical devices, the medical industry and equipment manufacturers lag behind in deployment of cybersecurity technology, and are lax when it comes to careful management of cybersecurity policies or procedures. Passwords used by medical staff are often weak or shared, or the default passwords that come pre-installed in medical equipment are unchangeable. Encryption is often nonexistent for data transmission between devices, and hacking of one device could enable unauthorized access to a host of others that are interconnected [4]. The demand for new portable medical devices will skyrocket as they become part of the Smart Grid networks that support the "Internet of Things." Device manufacturers need to find ways to improve the cybersecurity of their products.

## 3.3   SCADA Systems

SCADA Acronym for *supervisory control and data acquisition*, a computer system for gathering and analyzing real time data. SCADA systems are used to monitor and control a plant or equipment in industries such as telecommunications, water and waste control, energy, oil and gas refining, and transportation. It is a type of industrial control system (ICS). Industrial control systems are computer-based systems that monitor and control industrial processes that exist in the physical world. SCADA systems historically distinguish themselves from other ICS systems by being large-scale processes that can include multiple sites, and large distances.

SCADA systems have traditionally used combinations of radio and direct serial or modem connections for communications. The remote management or monitoring function of a SCADA system is often referred to as telemetry.

## 3.4 Smart Grid

The smart grid is considered the next stage of evolution for electric power utilities which opens a new dimension for civilian electric utilities and the national critical infrastructure. The smart grid is a modern electric power grid designed for improved efficiency, reliability, and safety, with smooth integration of renewable and alternative energy sources through automated control and modern communications technologies. While the traditional civilian electrical grid has been aging, the U.S. Department of Energy has reported that the demand and consumption for electricity in the U.S. have both increased by 2.5% annually over the last twenty years [5]. Consequently, a new grid infrastructure is urgently needed to address these challenges. The current electric power distribution network is not well suited to the needs of the twenty-first century. The new Smart Grid electric infrastructure is dynamic and is supported by two-way communications flowing between energy generation facilities, transmission devices for distribution, and end user consumption.

Today's critical electrical infrastructure has remained unchanged for almost one hundred years. Among the deficiencies are a lack of automated analysis, poor visibility of instantaneous demand and usage, and mechanical switches causing slow response times. These have contributed to several reported blackouts over the past 40 years. Additional factors include energy storage problems, the capacity limitations of current electricity generation, one-way communication, and the increase in prices for fossil fuels.

The Smart grid supports distributed power generation by renewable sources, and is intended to increase the efficiency, reliability, and safety of the existing power grid. A highly critical networked communications system is the key component of the smart grid infrastructure. The smart power grid infrastructure relies on sensors and highly critical communication paths to provide interoperability between distribution, transmission, and numerous substations, which includes residential, commercial, and industrial sites. This intelligent monitoring and control of power flowing between substations is enabled by modern digital communications technologies, many of which are HCWN [5].

## References

1. Hissam, J. P. (2013, Apr 7). *IEEE Explore, IEEE Wireless Communicat*. Retrieved from QoS optimization in ad hoc wireless networks through adaptive control of marginal utility: http://ieeexplore.ieee.org/xpl/articleDetails.jsp?reload=true&arnumber=655473

2. *Terrestrial Trunked Radio*. (2016, July 28). Retrieved from Wikipedia: https://en.wikipedia. org/wiki/Terrestrial_Trunked_Radio
3. Woodward, P. A. (2015, Jul 20). *Cybersecurity vulnerabilities in medical devices: a complex environment and multifaceted problem*. Retrieved from Medical Devices: http://www.ncbi.nlm. nih.gov/pmc/articles/PMC4516335
4. Bonderud, D. (2015, May 07). *Do No Harm? Medical Device Vulnerabilities Put Patients at Risk*. Retrieved from Security Intelligence: https://securityintelligence.com/news/do-no-harm-medical-device-vulnerabilities-put-patients-at-risk/
5. Gungor, V. C. (2011). *Smart Grid Technologies: Communication Technologies and Standards*. Retrieved from http://home.agh.edu.pl/~afirlit/LabRSMSM/Wykl045%20Smart%20Grids% 20-%20communication%20Technologies.pdf

# Chapter 4
# Vulnerabilities and Security Issues

HCWN networks provide flexibility for using mobile devices, and can be designed to increase reliability of communications in remote areas. However, the interdependencies and complexities of wireless systems in all industries are subject to cybersecurity vulnerabilities. These interdependencies are part of the designs that permit increased speeds and conveniences to help staff work faster and with more accuracy. Some examples where critical wireless technology improves worker performance are

1. Data links from programmable logic controllers in remote electrical substations and gas and water storage facilities permit better resource management.
2. Wireless smart meters permit remote reading of electric and gas meters, reducing distribution costs.
3. Smart grid communication more easily manages electricity distribution, load and bidirectional energy flow, and management of Internet-of-Things devices.
4. Wireless tablet-based terminals help staff isolate and manage faults in generating plants.
5. Production and distribution wireless networks enable remote management for oil and gas production, transportation, and truck tracking and scheduling.
6. Smartphone applications enable controls for home thermostats, security systems, and electronic keys for physical security.

Digital communications networks are vulnerable to several types of attacks, such as spoofing of Voice over IP (VoIP) automatic location identification records, as well as interception or corruption of the session initiation protocol records. Other security issues include abuse of dual-tone multifrequency (DTMF) tones, harvesting first responder pager and short message service (SMS) numbers, exploitation of real-time location data for field units, harvesting information about vulnerable citizens, and the possibility of publishing compromised data during lawsuits [1].

The United States reportedly is the most hacked country in the world. More than one quarter of all hacking attempts are directed at disrupting the data or the communications of the government and energy sectors. The communications industry is

© The Author(s) 2017
M. Martellini et al., *Information Security of Highly Critical Wireless Networks*,
SpringerBriefs in Computer Science, DOI 10.1007/978-3-319-52905-9_4

one of the next most popular targets for hacking. Once a communications network has been hacked, it takes an average of 205 days to detect the hack, according to research [1]. In more than two-thirds of hacking cases, the breach is discovered an external researcher, rather than being discovered by internal IT staff.

## 4.1    Wireless Vulnerabilities

Physical security of any wireless medium is impossible. This vulnerability enables hackers to eavesdrop or monitor wireless traffic not intended for them by setting their network adapters to "promiscuous" mode. This leaves the nodes of such a network vulnerable to loss of confidentiality, and also to "man-in-the-middle" attacks, wherein a hacker eavesdrops on messages between two or more nodes and relays or modifies messages so that the legitimate nodes are deceived into thinking they are talking directly to one another.

### *4.1.1    Wireless Eavesdropping*

Headsets, wireless phones, and wireless microphones often transmit messages "in the clear," which makes eavesdropping easy with inexpensive scanners such as KeyKeriki. Even if the attacker is only intercepting daily office conversation, victims are providing the attacker with information that can be used later for social engineering techniques. Digital communications reduce this particular type of vulnerability using encryption. However, digital communications has other types of vulnerabilities.

### *4.1.2    WEP and WPA Encryption*

WEP (Wired Equivalent Privacy) is a security encryption algorithm that was officially designated by IEEE in 2004 as too weak for adequate protection of network communications. However, many legacy wireless systems and equipment still depend on this weak security. Very little technical skill is required for an unauthorized user, or hacker, to discover the secret WEP network key in minutes using freely available software, and gain unauthorized access to the network. As a solution for this problem, IEEE released Wi-Fi Protected Access (WPA) wireless security encryption technology. The newest version, WPA2, employs the Advanced Encryption Standard (AES), but it only works with newer generation access points. Network systems that rely on older, legacy access points cannot take advantage of its improved security.

### *4.1.3 Jamming*

All wireless radio emissions are vulnerable to jamming, which is a method to deliberately overpower or disrupt legitimate broadcast signals. For instance, a global positioning system (GPS) jammer can be constructed for around $30 with equipment obtainable from most electronic supply stores. An inexpensive jammer, such as this, can overpower GPS signals within a 75-mile radius. By submitting multiple phony authentication requests to an access point, an attacker can overwhelm its processing resources, preventing legitimate clients from connecting though the access points.

### *4.1.4 Rogue Access Points*

Many wireless digital circuit cards have the ability to operate as a wireless access point. However, it is easy for a hacker to impersonate a legitimate Access Point (AP) by simply copying the Service Set Identifier (SSID) for the circuit card. This is because the 802.11 communication standard for authentication is one-way, from access point to client. Clients could be fooled into connecting to a rogue access point.

### *4.1.5 Injection Attacks*

Hackers can eavesdrop on legitimate traffic with a freely available digital packet sniffer. If the access point is open, it is easy to quickly read and reply to a message with a fake reply. With freely available packet injection software like Airpwn, an unauthorized user can send modified versions of legitimate requests before the authentic web server has a chance to respond. When the legitimate reply arrives a moment later, it may be rejected by the client as erroneous.

## 4.2 Medical Device Vulnerabilities

Software is embedded into all digital medical device to assist in operation and accuracy. Well-developed and validated software has the potential to significantly and positively affect the delivery of patient care, transforming how we manage health care across the globe. However, the exposure of devices to networking has increased the risk for cyberattack. Medical devices that are no longer a stand-alone, such as implantable medical devices capable of being reprogrammed wirelessly, are now vulnerable to cyberattack. Examples include pacemakers, drug (e.g., insulin) pumps, defibrillators, and neuro-stimulators that are now used for monitoring and treating patients.

A few common vulnerabilities that can be found in digital wireless devices used in the hospital environment include web interfaces to infusion pumps, default hard coded administration passwords, and possible access to the external Internet through connected internal networks. Embedded web services, with unauthenticated and unencrypted communication are one of the biggest vulnerabilities, as an attacker can potentially affect these devices remotely from anywhere in the world. More than 2.5 million implantable medical devices are currently in use, and that number is expected to grow almost 8% this year. The Industrial Control Systems Cyber Emergency Response Team (ICS-CERT) reportedly found that 300 devices have unchangeable passwords. If malicious actors ever obtained a complete list of these static passwords, there would be no way to prevent misuse short of tossing the device in the garbage [1], [2].

For networked medical devices and mobile health technologies, these types of vulnerabilities may expose patients and healthcare organizations to safety and security risks that are life threatening. Networking also increases the risk for access by users who are unauthorized. Medical devices that incorporate wireless capabilities and complex software are eventually connected to traditional wired medical devices in hospitals, health systems, and home-based systems. This causes the scope and nature of required security controls to also change [3]. Healthcare organizations will need to anticipate present and future medical device security risks to safeguard patient safety and protect medical records.

## 4.3 Smart Grid, Mesh Network Vulnerabilities

The smart grid, generally referred to as the next-generation power system, is considered the next evolution of the current massive, regional power grids. However, potential network intrusion by adversaries who attack smart grid technologies and equipment may lead to of severe consequences such as customer information leakage, or a cascade of failures, such as massive blackout and destruction of infrastructures [4].

Two-way metering for smart grid systems and the "Internet-of-Things" will essentially turn every single household appliance into the equivalent of a digital transmitting cell phone. That includes every dishwasher, microwave oven, stove, washing machine, clothes dryer, air conditioner, furnace, refrigerator, freezer, coffee maker, TV, computer, printer, and fax machine. The average U.S. home has over 15 such appliances, and as smart grid is implemented, each device would be equipped with a transmitting antenna for two-way communication with the power generator. General Electric (GE) and other appliance manufacturers are already putting transmitters into their latest product designs, and the U.S. Department of Energy (DOE) is already providing tax credits to manufacturers. All transmitters inside the home or office will communicate with a Smart Meter, or house meter, attached outside each building, and then, using a higher frequency, the Smart Meter would communicate with a central hub installed in local neighborhoods. In what are called "mesh networks," signals can also be bounced from house meter to house meter before reaching the final central hub [5].

The smart grid meters and antennas for new interconnected household appliance will act as transceivers. This enables the customer to control individual home appliances remotely and wirelessly. Also, the utility company can transmit signals to remotely control all individual appliances wirelessly. Reportedly, one such system in operation in the Midwest already allows the local utility to cycle household furnaces and air conditioners on and off every 15 min, with the stated purpose to reduce peak-loads on electric grids. However, this introduces a new vulnerability. As home energy use can be recorded in real time, it is easy to determine when a customer is present in the home, or away from home [5]. By monitoring usage data, or simple monitoring of the energy spikes in data transmission, a hacker can eavesdrop by setting their network equipment to "promiscuous" mode, or by setting up rogue access points.

# References

1. Beckman, K. (2015, June 01). *Mission-Critical Networks Face Increasing Cybersecurity Threats*. Retrieved from Spectrum Monitor: http://digital.olivesoftware.com/Olive/ODN/MissionCritical/PrintArticle.aspx?doc=MCR%2F2015%2F06%2F01&entity=ar01400
2. Woodward, P. A. (2015, Jul 20). *Cybersecurity vulnerabilities in medical devices: a complex environment and multifaceted problem*. Retrieved from Medical Devices: http://www.ncbi.nlm.nih.gov/pmc/articles/PMC4516335
3. Bonderud, D. (2015, May 07). *Do No Harm? Medical Device Vulnerabilities Put Patients at Risk*. Retrieved from Security Intelligence: https://securityintelligence.com/news/do-no-harm-medical-device-vulnerabilities-put-patients-at-risk/
4. Lu, W. W. (2012, April 06). *Cyber security in the Smart Grid: Survey and challenges*. Retrieved from Computer Networks Journal: http://www.ece.ncsu.edu/netwis/papers/13wl-comnet.pdf
5. Glendenning, C. (2011, Mar 18). *The Problems with Smart Grids*. Retrieved from Counterpunch: http://www.counterpunch.org/2011/03/18/the-problems-with-smart-grids/

# Chapter 5
# Modeling Threats and Risks

Scanning of critical infrastructure networks is done for intelligence gathering for industrial espionage or for making preparations to direct a future cyberattack. Trend Micro officials reportedly have stated that, "We have observed increased interest in [scanning of] SCADA protocols …" for critical infrastructure and industrial systems [1]. Hackers use cyber espionage to read network traffic and then gain access to the network to install malware. In many cases, backdoors are installed into the system enabling easier access later. Many tools and files used to gain unauthorized access to the network are also self-erasing [1].

## 5.1 Passive Attacks

Most passive attacks on wireless networks involve an attacker with unauthorized access to the wireless link. In passive attacks, the attacker monitors network communications for data, including authentication credentials or transmissions that identify communication patterns and participants. Information is collected, such as stolen passwords, and can be later used for follow-on attacks, such as impersonating a legitimate user. Passive attacks can occur at any point in the wireless network.

## 5.2 Active Attacks

Active attacks rely on an attacker's ability to intercept and inject false information directly into network transmissions. Message can be deleted or changed this way. Without the use of encryption, wireless transmissions can be intercepted and easily monitored or copied by anyone within range. Although it is not the typical case, an

© The Author(s) 2017
M. Martellini et al., *Information Security of Highly Critical Wireless Networks*,
SpringerBriefs in Computer Science, DOI 10.1007/978-3-319-52905-9_5

intercepting receiver can receive the target signals because standardized commercial communication protocols are readily available [2]. When communications are jammed, the legitimate message is disrupted or overpowered with a stronger radio signal.

## References

1. Beckman, K. (2015, June 01). *Mission-Critical Networks Face Increasing Cybersecurity Threats*. Retrieved from Spectrum Monitor: http://digital.olivesoftware.com/Olive/ODN/ MissionCritical/PrintArticle.aspx?doc=MCR%2F2015%2F06%2F01&entity=ar01400
2. Ewing, M. H. (2010, Nov 7). *Wireless Network Security in Nuclear Facilities*. Retrieved from United States Nuclear Regulatory Commission: http://www.nrc.gov/docs/ML1032/ ML103210371.pdf

# Chapter 6
# Modeling Vulnerabilities

The Check Point Software organization releases a security report in 2015 which found that mobile communications devices have become the biggest threat for today's enterprises [1]. For example, the report showed that 82% of businesses now have some kind of plan in place where employees are allowed to use their personal wireless devices at work. In many cases, employees are allowed to connect their personal devices to the organization's corporate network. This phenomenon is called "Bring Your Own Device" to work, or BYOD. Even heavily regulated industries like healthcare and financial services are putting BYOD programs in place because of pressure from the lines of business. Today, businesses that do not allow workers to use mobile devices are putting themselves at a competitive disadvantage. The Check Point survey also found that organizations with more than 2000 devices on the network have a 50% chance that at least six of them are infected [1].

Without an adequate security policy in place, supplemented by proper monitoring of worker activity and network traffic, this can expose the organization to a variety of cybersecurity issues. For example, if a wireless router were to be plugged into an organization's unsecured switch port, the entire network can be exposed to anyone within range of the signals. Similarly, if an employee adds a wireless interface to a networked computer using an open USB port, they may create a breach in network security that would allow unauthorized access to confidential materials [2]. Non-traditional networks such as personal network Bluetooth devices are not safe from hacking and should also be regarded as a security risk. Even Barcode readers, handheld personal data assistants, and wireless printers should be secured.

© The Author(s) 2017
M. Martellini et al., *Information Security of Highly Critical Wireless Networks*,
SpringerBriefs in Computer Science, DOI 10.1007/978-3-319-52905-9_6

# References

1. Kerravala, Z. (2015, Aug 23). *Mobile devices pose biggest cybersecurity threat to the enterprise, report says*. Retrieved from Network World: http://www.networkworld.com/article/2974702/cisco-subnet/mobile-devices-pose-biggest-cybersecurity-threat-enterprise-report.html
2. *Wireless Security*. (2016, Aug 02). Retrieved from Wikipedia: https://en.wikipedia.org/wiki/Wireless_security

# Chapter 7
# Governance and Management Frameworks

Wireless networks are governed by a variety of regulations that are intended to cover safe operation of emergency communications and medical devices.

## 7.1 FCC Rules

The nation's communication infrastructure is becoming increasingly complex, and there have been recent outages for the 9-1-1 emergency phone number caused by software and database errors. The FCC has proposed rules to keep communications reliable for TCP–IP networking protocols [1].

The Industry Council for Emergency Response Technologies (iCERT) has proposed four policy statements in 2015 to improve cybersecurity for communications:

1. Public policies should promote increased reliability and resiliency of 9-1-1 systems.
2. A transition to new technology should not result in a loss of 9-1-1 location accuracy capability.
3. The transition to new technologies has shifted some of the responsibility for providing backup power to the consumer, especially for devices.
4. The transition to IP-based networks introduces new risks related to cybersecurity. Public safety, service providers, and consumers each have a role to play in cybersecurity [1].

## 7.2 Spectrum Sharing

In 2014, the federal government adopted spectrum sharing by adopting report from the President's Council of Advisors on Science and Technology (PCAST). It is believed that next generations of wireless communications technology will embed spectrum sharing as part of their protocols.

© The Author(s) 2017

M. Martellini et al., *Information Security of Highly Critical Wireless Networks*, SpringerBriefs in Computer Science, DOI 10.1007/978-3-319-52905-9_7

## 7.3  FDA

Regulatory authorities, such as the US Food and Drug Administration (FDA), have responsibility for assuring the safety, effectiveness, and security of medical devices. The regulatory bodies have acknowledged the seriousness and enormity of the problem by publishing recommendations for managing cybersecurity risks and protecting patient health information, to assist manufacturers in their submissions for FDA approval of medical devices [2]. In addition, because thumb drives are a major potential source of virus infections in medical devices, "we scan portable media before it's connected to a device," Friel says [3].

The FDA has ruled, under the Medical Device Data System Rule, that medical device regulation includes "software, electronic and electrical hardware, including wireless." In addition, this rule by FDA also includes data storage and data transfer, which has not been a security focus for medical device manufacturers [2].

On January 15, 2016, the Food and Drug Administration issued the "Draft Guidance for Industry and Food and Drug Administration Staff," advising medical device manufacturers to address cybersecurity "throughout a product's lifecycle, including during the design, development, production, distribution, deployment, and maintenance of the device."

The guidelines are voluntary, and show how organizations can ensure that their cybersecurity policies, procedures, and strategies proactively address cybersecurity risks in medical devices before there is harm from the exploitation of an unaddressed vulnerability by an unknown threat actor. The draft guidelines start with NIST's 2014 "Framework for Improving Critical Infrastructure Cybersecurity," which in turn was published in response to President Obama's Executive Order 13636 that advocates the development of a standardized cybersecurity framework that identifies, detects, protects against, responds, and recovers from cybersecurity risk [4].

Regulatory frameworks are difficult to develop and enforce because different organizations operate under different constraints. Regulations are developed as bare minimums, inadequate to the actual threat, because the regulatory body can only enforce according to the maximum capability of the weakest organization [4].

## References

1. Communications, M. (2015, Jan 21). *iCERT Releases Policy Statement on 9-1-1 Technology Transition (1/21/15).* Retrieved from RadioSource International: http://www.rrmediagroup. com/News/NewsDetails/NewsID/11629/
2. Woodward, P. A. (2015, Jul 20). *Cybersecurity vulnerabilities in medical devices: a complex environment and multifaceted problem.* Retrieved from Medical Devices: http://www.ncbi.nlm. nih.gov/pmc/articles/PMC4516335/

3. Anderson, H. (2011, May 17). *Medical Device Security Raises Concerns*. Retrieved from Healthcare Info Security: http://www.healthcareinfosecurity.com/medical-device-security-raises-concerns-a-3644
4. Scott, J. (2016). *Assessing the FDAs Cybersecurity Guidelines for Medical Device Manufacturers*. Retrieved from Institute for Critical Infrastructure Technology: http://icitech.org/wp-content/uploads/2016/02/ICIT-Blog-FDA-Cyber-Security-Guidelines2.pdf

# Chapter 8
# Security Technologies for Networked Devices

## 8.1 Basic Security Controls for All Wireless Networks

   I. To protect against "man-in-the-middle" attacks, wherein a hacker eavesdrops on messages between two or more nodes and relays or modifies messages so that the legitimate nodes are deceived into thinking they are talking directly to one another. To protect from such attacks, a successful information assurance strategy must make the mere reception of a signal useless to a would-be hacker.

  II. Encryption: using cryptography to encrypt wireless communications prevents exposure of data through eavesdropping.

 III. Cryptographic Hashes: calculating cryptographic hashes for wireless communications allows the device receiving the communications to verify that the received communications have not been altered in transit, either intentionally or unintentionally. This prevents masquerading and message modification attacks.

  IV. Device Authentication: authenticating wireless endpoints to each other prevents man-in-the-middle attacks and masquerading.

   V. Replay Protection: adding devices such as incrementing counters, timestamps, and other temporal data to communications to detect message replay.

  VI. Wireless Intrusion Detection: monitoring events in network or computer systems and analyzing them for a possible violation of the network security or simple standard policies.

 VII. Physical Security: limiting physical access within the range of the wireless network to prevent some jamming and flooding attacks [1].

© The Author(s) 2017
M. Martellini et al., *Information Security of Highly Critical Wireless Networks*,
SpringerBriefs in Computer Science, DOI 10.1007/978-3-319-52905-9_8

## 8.2  Encryption

Encryption is most commonly used to protect against unauthorized access or damage to computers using wireless networks. The most common types of wireless security are Wired Equivalent Privacy (WEP) and WiFi Protected Access (WPA). WEP is a notoriously weak security standard, and the password it uses can often be cracked in a few minutes with a basic laptop computer and widely available software tools. It has been replaced by WPA, or Wi-Fi Protected Access, which offers stronger security. A longer encryption key length improves security over WEP. The current standard is WPA2, but some network equipment cannot support WPA2 without a firmware upgrade or replacement [2].

Messages sent over wireless links must be encrypted to maintain confidentiality. Because information transmitted with weaker or no encryption is vulnerable to interception by an intruder, and the origin of messages received over wireless links must be verified for authenticity [1]. Network Access Control (NAC) enables security by registering all devices connected to a network. NAC allows the network administrator to know who is connected to the network, what device is being used, what applications open.

## 8.3  Directional Transmission and Low Power Signals

Another method that can be used for added security is directional transmission of wireless signals. This can be enabled using a focusing antenna, or dish. Transmitting the data directly towards the intended receiver reduces the locations from which the transmissions may be received. If this method is combined with low power signals, it can be even more effective. A further optimization of this technique could involve multiple access points utilizing phased array antennas. The signal can be multiplexed between the access points so that parts of the signal are transmitted from each access point directly toward the receiver. In this way, an eavesdropper would not be able to intercept the entire signal without having at least one antenna located in line with each transmitter and receiver [1].

## References

1. Ewing, M. H. (2010, Nov 7). *Wireless Network Security in Nuclear Facilities*. Retrieved from United States Nuclear Regulatory Commission: http://www.nrc.gov/docs/ML1032/ML103210371.pdf
2. *Wireless Security*. (2016, Aug 02). Retrieved from Wikipedia: https://en.wikipedia.org/wiki/Wireless_security

# Chapter 9
# Known Weaknesses with Security Controls

Richard Clarke, former White House advisor on cybersecurity, has reportedly warned that there is evidence that China has been actively probing and hacking wireless networks that control the United States power grid. Clarke points out, "The only point to penetrating the grid's controls is to counter American military superiority by threatening to damage the underpinning of the U.S. economy. Chinese military strategists have written about how in this way a nation like China could gain an equal footing with the militarily superior United States" [1].

There are over 2.9 million active very small aperture terminals (a VSAT terminal is a two-way satellite ground station with a dish antenna that is smaller than 3 m—sometimes used for point-of-sale transactions using credit cards, or for SCADA system control) in the world, with two-thirds of those devices the U.S., being used by US defense contractors, or the military, to transmit government and classified communications. Others are used by financial industries like banks to transmit sensitive data, and still others are used by the industrial sector such as energy to transmit from power grid substations, or oil and gas to transmit from oil rigs. After running a scan, Cyber intelligence firm IntelCrawler, found that over 10,000 of those devices are reportedly "open" for targeted cyberattacks. Reportedly, many of the "VSAT devices have telnet access with very poor password strength, many times using default factory settings. The fact that one can scan these devices globally and find holes is similar to credit card thieves in the early 2000s just goggling the terms 'order.txt' and finding merchant orders with live credit cards" [2].

© The Author(s) 2017
M. Martellini et al., *Information Security of Highly Critical Wireless Networks*,
SpringerBriefs in Computer Science, DOI 10.1007/978-3-319-52905-9_9

# References

1. Bradley, T. (2011, Jun 17). *PC World*. Retrieved from SCADA Systems: Achilles Heel of Critical Infrastructure: http://www.pcworld.com/article/230675/scada_systems_achilles_heel_of_critical_infrastructure.html
2. Scott, J. (2016). *Assessing the FDAs Cybersecurity Guidelines for Medical Device Manufacturers*. Retrieved from Institute for Critical Infrastructure Technology: http://icitech.org/wp-content/uploads/2016/02/ICIT-Blog-FDA-Cyber-Security-Guidelines2.pdf

# Chapter 10
# Competent Reliable Operation of HCWN

Telemetry is the automatic transmission and measurement of data from remote sources by wire or radio or other means. It is also used to send commands, programs and receives monitoring information from these terminal locations. Supervisory Control and Data Acquisition (SCADA) systems use a combination of telemetry and data acquisition. SCADA is used for collecting information, transferring it to a central management site, carrying out any necessary data analysis and control operations, and then displaying the status information back onto the operator screens. Typically, SCADA systems include the following components:

1. Operating equipment such as pumps, valves, conveyors, and substation breakers that can be controlled by energizing actuators or relays.
2. Instruments in the field or in a facility that sense conditions such as pH, temperature, pressure, power level, and flow rate.
3. Local processors that communicate with the site's instruments and operating equipment. This includes the Programmable Logic Controller (PLC), Remote Terminal Unit (RTU), Intelligent Electronic Device (IED) and Process Automation Controller (PAC).
4. A single local processor may be responsible for dozens of inputs from instruments and outputs to operating equipment.
5. Short range communications between the local processors and the instruments and operating equipment.
6. These relatively short cables or wireless connections carry analog and discrete signals using electrical characteristics such as voltage and current, or using other established industrial communications protocols.
7. Host computers that act as the central point of monitoring and control. The host computer is where a human operator can supervise the process; receive alarms, review data, and exercise control.
8. Long range communications between the local processors and host computers. This communication typically covers miles using methods such as leased phone lines, satellite, microwave, frame relay, and cellular packet data [1].

© The Author(s) 2017
M. Martellini et al., *Information Security of Highly Critical Wireless Networks*,
SpringerBriefs in Computer Science, DOI 10.1007/978-3-319-52905-9_10

# Reference

1. Kim, T.-h. (2010, Vol 4). *Integration of Wireless SCADA through the Internet*. Retrieved from INTERNATIONAL JOURNAL of Computers and Communications: http://www.universitypress.org.uk/journals/cc/19-833.pdf

# Chapter 11
# Assessing the Effectiveness and Efficiency of Security Approaches

## 11.1 WEP Legacy Issues

As technology has evolved, Wired Equivalent Privacy (WEP) protocol is no longer considered effective for establishing a secure wireless network. Today, the tools found in common penetration testing kits are now fully automated, with GUIs that make cracking a WEP key as easy as point and click. Once it became apparent that WEP had fatal, unfixable flaws, there were immediate efforts to develop a successor. Since a replacement was needed immediately, there was an interim standard developed called WiFi Protected Access (WPA) published in 2003, which was further refined as WPA2 in 2004 [1]. With more secure alternatives on the market for over nine years, it seems like WEP would be all but extinct, but sadly that is not the case. WEP still remains in use in many places.

## 11.2 Use of a DMZ for SCADA

Until recently, SCADA systems which monitored and operated facilities on the shop floor, or at remote locations, were usually not connected directly to any network or wireless system. This physical separation offered a measure of protection against cybersecurity vulnerabilities that might affect the corporate network. However, as technology evolved, corporate networks and wireless systems have been connected to most SCADA equipment at remote locations to facilitate more rapid control and monitoring from within the corporate network. This network connection and wireless linkage has exposed SCADA systems to increased vulnerability to cyberattack.

© The Author(s) 2017
M. Martellini et al., *Information Security of Highly Critical Wireless Networks*,
SpringerBriefs in Computer Science, DOI 10.1007/978-3-319-52905-9_11

However, to increase protection against cyberattack, a network with a SCADA system can be segmented to create an architecture with security zones that provide access control by separating systems with different security and access require- ments. A DMZ (De-Militarized Zone, or perimeter subnetwork) architecture pro- vides this separation, where the ICS network (Internet Connection Sharing (ICS) is any device with Internet access, or Internet gateway) is separated from other por- tions of the corporate network by multiple firewalls. The DMZ should provide the corporate network access to the required information from the ICS network or SCADA system. A virtual private network (VPN) can be used to enable encrypted connections between the ICS and other portions of the corporate network for acceptable communications. Only restricted, encrypted communication should occur between the corporate network and the DMZ, and the ICS network and the DMZ. The corporate network and the ICS network should not communicate directly with each other (NIST 800-41) [2].

Creating architecture for a DMZ requires that the firewalls used offer three or more interfaces, rather than the typical public and private interfaces. One connected to the corporate network, another to the control network, and the remaining interfaces to the shared or insecure devices such as the data historian server or wireless access points on the DMZ network. No direct communication paths are allowed from the corporate network to the control network; each path effectively ends in the DMZ. Most firewalls can allow for multiple DMZs, and can specify what type of traffic may be forwarded between zones. The firewall can block arbitrary packets from the corporate network from entering the control network and can regulate traffic from the other network zones including the control network. With well-planned rule sets, a clear separation can be maintained between the control network and other networks, with little or no traffic passing directly between the corporate and control networks. The primary security risk in this type of architecture is that if a computer in the DMZ is compromised, it can be used to launch an attack against the control network via application traffic permitted from the DMZ to the control network [2].

# References

1. Tokuyoshi, B. (2013, Aug 26). *Diving into Wireless Network Threats – Weaknesses in WEP*. Retrieved from Paloalto: http://researchcenter.paloaltonetworks.com/2013/08/diving-into-wireless-networks-threats-weaknesses-in-wep/
2. DHS. (2011, May 01). *Common Cybersecurity Vulnerabilities in Industrial Control Systems*. Retrieved from National Cyber Security: https://ics-cert.us-cert.gov/sites/default/files/recommended_practices/DHS_Common_Cybersecurity_Vulnerabilities_ICS_2010.pdf

# Chapter 12
# Examples in Brief

Below are a few examples where vulnerabilities and threats to HCWN are described.

## 12.1  SCADA Software from China

According to a warning issued by the U.S. Industrial Control Systems Cyber Emergency Response Team (ICS-CERT), two vulnerabilities found in industrial control system software made in China but used worldwide could be remotely exploited by attackers. The vulnerabilities were found in two products from "Sunway ForceControl Technology," a maker of software for a wide variety of industries, including defense, petrochemical, energy, water and manufacturing, the agency said. Sunway's products are used in Europe, the Americas, Asia, and Africa. According to the warning by US iCERT, the problems could supplement a denial of service cyberattack. Both issues were found by Dillon Beresford, who works for the security testing company NSS Labs [1].

## 12.2  Angen 9-1-1

In 2015, the Alabama 9-1-1 Telephone Board for emergency communications completed a study of cybersecurity risks to the Alabama Next Generation Emergency Network (ANGEN) [2]. During the study of its network vulnerabilities, ANGEN experienced an outside breach. The system is highly integrated with other government systems, including public schools and universities, said Jackson. A college student trying to hack into the school's system eventually gained access to the 9-1-1 network and attempted to launch a DDoS attack. In response, ANGEN increased its firewalls, updated all of its routers and has a plan to isolate breaches and shut down affected PSAPs until the problem can be solved, said Jackson [2].

© The Author(s) 2017                                                                                  33
M. Martellini et al., *Information Security of Highly Critical Wireless Networks*,
SpringerBriefs in Computer Science, DOI 10.1007/978-3-319-52905-9_12

## 12.3   General Dynamics Smartphones

In 2012, General Dynamics C4 Systems announced that it will integrate its defense-grade cyber and information security technologies into the line of Samsung Approved for Enterprise (SAFE) smartphones and tablet computers. The devices will also be upgraded with the Samsung Secure Android platform and are designed for customers ranging from the military and government agencies to law enforcement and public safety to utility and other industry personnel who need to browse the public Internet and access secure networks [3].

General Dynamics C4 Systems delivered 300 rugged smartphones to the U.S. Air Force, for use by senior leadership at the air staff and major command levels. The smartphones are part of a broader Air Force plan to integrate secure mobile devices into its consolidated enterprise network. The Sectéra Edge is certified by the National Security Agency for classified voice and data, using wireless access to commercial WiFi and cellular networks that provide access to classified and unclassified government networks. The device is capable of synchronizing information with a user's computer, enabling access to calendar, address book, calculator, notepad and other PDA capabilities [4].

## 12.4   Medical Devices at VA

Cybersecurity vulnerabilities for medical devices consists of multiple factors. These include the transfer from isolated devices to networked, and the tensions this creates between security and safety. Cybersecurity is not just a technical problem [5].

In January 2009, the US Department of Veterans Affairs discovered that up to 173 medical devices had been infected with malware. In response, the VA isolated 50,000 medical devices behind nearly 3200 virtual local area networks to improve security [6]. Meanwhile, the VA is validating its new virtual local area networks for medical devices that it recently spent seven months installing to help improve security. It is also using ACLs, or access control lists, that, among other things, prevent linking devices to the Internet [6]. However, this security control defeated the interoperability and connectivity required by mobile medical devices [5].

## 12.5   Drug Infusion Pump

Recently, a medical device known as the Hospira PCA3 Drug Infusion pump was reportedly found to have cybersecurity problems. The device could be completely disabled during operation if the operator made a single error while typing instructions. The pumps all used a default IP address of 192.168.0.100, a familiar address to hackers who could access wireless encryption keys from the medical device, which were stored in plaintext [7].

Reportedly, the pumps do not require authentication for their drug libraries, meaning anyone with access to the hospital's network could potentially load up a new drug library and change the dosage being administered. Altering the upper or lower dosage limits could be fatal to a hospital patient. Reports of these vulnerabilities have been passed on to manufacturer of the Hospira PCA3 Drug Infusion pump, but so far the company has been quiet on any potential fix [7].

# References

1. Kirk, J. (2011, Jun 17). *PC World*. Retrieved from US Warns of Problems in Chinese SCADA Software: http://www.pcworld.com/article/230530/article.html
2. Beckman, K. (2015, June 01). *Mission-Critical Networks Face Increasing Cybersecurity Threats*. Retrieved from Spectrum Monitor: http://digital.olivesoftware.com/Olive/ODN/MissionCritical/PrintArticle.aspx?doc=MCR%2F2015%2F06%2F01&entity=ar01400
3. Communications, M. (2012, May 23). *General Dynamics Adds Security to Samsung Devices for Public Safety (5/23/12)*. Retrieved from RadioResource International: http://www.rrmediagroup.com/News/NewsDetails/NewsID/8299/
4. Communicaitons, M. (2011, June 16). *General Dynamics Delivers 300 Secure Smartphones to Air Force (6/16/11)*. Retrieved from RadioSource International: http://www.rrmediagroup.com/News/NewsDetails/NewsID/7134/
5. Woodward, P. A. (2015, Jul 20). *Cybersecurity vulnerabilities in medical devices: a complex environment and multifaceted problem*. Retrieved from Medical Devices: http://www.ncbi.nlm.nih.gov/pmc/articles/PMC4516335
6. Anderson, H. (2011, May 17). *Medical Device Security Raises Concerns*. Retrieved from Healthcare Info Security: http://www.healthcareinfosecurity.com/medical-device-security-raises-concerns-a-3644
7. Bonderud, D. (2015, May 07). *Do No Harm? Medical Device Vulnerabilities Put Patients at Risk*. Retrieved from Security Intelligence: https://securityintelligence.com/news/do-no-harm-medical-device-vulnerabilities-put-patients-at-risk/

# Chapter 13
# Testing the Resilience of HCWN

## 13.1   Introduction

The wireless technologies, since their birth in 1890, have revolutionized our life style and provided freedom and liberty in remote data procession and management, both in civilian and military areas. After more than a century of evolution, wireless technology is adopted globally due to its advantages and ease of use in communication and data transfer, and wireless interconnectivity surrounds us on a daily basis, from wireless CCTV cameras to smartphones.

The new advantages are followed by the emerging cybersecurity challenges, and we enquire if "wireless" means "secured". Can information, potentially available to anyone with a configured receiver, be properly protected? And what is the cyber secured future of wireless technologies?

Information, being transferred over long distances and literally in the air, might seem unprotected and encourages attackers to "sniff" or directly connect to wireless access points, retrieve and impact organizations data confidentiality, integrity, authentication and access control. In addition, interdependencies and complexities of industrial and corporate wireless systems, designed to speed up the work efficiency, generate cyber security vulnerabilities by default.

Numerous organizations develop today wireless technology security and standards. Among them, there are ISO,[1] ITU-T,[2] IEEE,[3] ETSI,[4] NCITS,[5] as well as alliances such as the WiFi Alliance[6] and WiMax Forum[7] which do research in computer networks, including in wireless security.

---

[1]http://www.iso.org/iso/home.
[2]http://www.itu.int/en/ITU-T/Pages/default.aspx.
[3]http://www.ieee.org/index.html.
[4]http://www.etsi.org/.
[5]http://www.ncits.org/.
[6]http://www.wi-fi.org/.
[7]http://wimaxforum.org/.

© The Author(s) 2017
M. Martellini et al., *Information Security of Highly Critical Wireless Networks*,
SpringerBriefs in Computer Science, DOI 10.1007/978-3-319-52905-9_13

Though basic security measures for wireless network have already been developed and successfully deployed, it is important to regularly upgrade them, monitor, and assess their implementation. Especially crucial it is for mission critical communications and military, medical and CBRNe infrastructure. As the more sophisticated the target and ramifications are, the more interest they evoke from attackers and the higher protection skills are requested from defenders. "The sheer number and complexity of potential targets guarantee that terrorists can find weaknesses and vulnerabilities to exploit" [1].

## 13.2   Definitions

Cyber security testing is a process intended to reveal flaws in the security mechanisms of an information system that protects data and maintains functionality as intended.[8] It is applied in a specific cyber environment and includes several components. The following definitions will be accepted for the purpose of the article.

*Wireless network* is any type of computer network that uses wireless data connections for connecting network nodes [2].

*Local Area Network* (*LAN*) is a computer network that interconnects computers within a limited area such as a residence, school, laboratory, university campus, or office building and has its network equipment and interconnects locally managed [3].

*Wireless Access Point* (*AP*) is a networking hardware device that allows a wireless protocol compliant device to connect to a wired network.

*Resilience of the system* is the ability to provide and maintain an acceptable level of service in the face of faults and challenges to normal operation.[9]

*Beacon frame* is one of the management frames in IEEE 802.11 based WLANs, which contains all information about the network. Beacon frames are transmitted periodically to announce the presence of a wireless LAN, and they are transmitted by the AP in an Infrastructure basic service set (IBSS). In IBSS network beacon generation is distributed among the stations.

*Authentication in wireless communication* is the process of self-identification between two devices (e.g., device and AP). At this stage no data encryption or security is used.

*Association in wireless communication* is the process of the device registration in the AP/router database. In association a single device can only associate with one AP/router at a time. At this stage encryption handshake can be configured.

*Encryption handshake* is a sequence of request/response packets between the device and AP to provide mutual authentication.

---

[8]National Information Assurance Glossary, https://www.ecs.csus.edu/csc/iac/cnssi_4009.pdf.
[9]https://wiki.ittc.ku.edu/resilinets_wiki/index.php/Definitions#Resilience.

## 13.3  Goals of Cyber Security Testing

According to a globally recognized model, developed by the National Institute of Standards and Technology [4], all information security measures try to address at least one of the three goals:

- Protect the confidentiality of data
- Preserve the integrity of data
- Promote the availability of data for authorized use.

Thus, the three security goals are "Confidentiality", "Integrity", and "Availability" (CIA), which form the CIA triad, the basis of all information security designs. The development of all information security policies and procedures is guided by these three goals.

To guarantee CIA of any computer system or network, cyber security testing detects and classifies flaws in cybersecurity. It identifies risks or the threats in the system and/or network; classifies potential vulnerabilities; and further helps in developing software, hardware, or physical solutions.

As the wireless systems have a more sophisticated design than the wired computer systems, their security testing is more complicated. In addition, it should acknowledge detectability and vulnerability of routers and adapters to further develop and deploy preventive measures against "eavesdropping", denial of service, security breaches, and unauthorized remote control of wireless devices.

## 13.4  Types of Cyber Security Testing

According to the pre-engagement agreement between the organization and the security testing team, (internal or external) the security test might be conducted with zero, partial, or full knowledge of the target system. Hence, the following types of penetration testing might be defined:

- white box testing—complete knowledge of the systems
- gray box testing—partial knowledge of the systems
- black box testing—no knowledge of the system or its credential.

*White box testing* refers to cyber security testing with full knowledge and access to all source code and documentation on network architecture. The open access to this sensitive information ensures high efficiency in revealing vulnerabilities and inconsistencies. The *white box testing* provides the maximum information about the internal and perimeter defenses, as well as about the level of exposure of the client to potential attackers.

*Gray box testing* means testing a system or network while having limited knowledge of the internal architecture. This information is usually constrained to detailed design documents and network architecture diagrams. It is a combination of both black and white box testing capacities.

*Black box testing* refers to testing a system without having any knowledge of the internal work of the system, with no access to the source code, and no knowledge of the architecture. This technique is the most sophisticated and resource consuming, it also requires a high level of expert knowledge.

## 13.5   Network Communication Standards

The classification of wireless communication networks is based on radio frequencies spectrum allocated and communication protocols implemented for data transmission. The international classification of networks is known as IEEE 802 standards. The IEEE stands for the Institute of Electrical and Electronics Engineers—a globally recognized technical professional organization that develops technological standards and innovations.[10]

IEEE 802 standards are a family of IEEE standards primarily implemented in local area networks and metropolitan area networks, including wireless communications, restricted to networks carrying variable size packets. The number 802 was the next free number IEEE could assign, though "802" is debated to be associated with the date the first meeting on network standards was held—1980, February [5].

Table 13.1 (Table of IEEE standards) enumerates network standards, developed by IEEE throughout the years.

Hardware and protocol implementations vary and depending on the deployed hardware and protocol implementations, the security testing approaches may change.

## 13.6   Wireless Networks by Geographical Range

The wireless architecture allows any device to connect in the radial proximity to the AP. The operation architecture of wireless networks usually incorporates a notion of scalability, including the possible expansion of the network. The network design assumes the possible maximum number of devices, connected at the same time, as well as the physical layout of the interconnected devices and access points. With the increased sophistication of the deployed network, misconfigurations of the equipment, vulnerabilities, as well as landscape and urban territory may provide advantages to the attacker.

---

[10]https://www.ieee.org/about/index.html.

**Table 13.1** Table of IEEE standards[a]

| Name | Description | Note |
|---|---|---|
| IEEE 802.1 | Higher Layer LAN Protocols | Active |
| IEEE 802.2 | LLC | Disbanded |
| IEEE 802.3 | Ethernet | Active |
| IEEE 802.4 | Token bus | Disbanded |
| IEEE 802.5 | Token ring MAC layer | Disbanded |
| IEEE 802.6 | MANs (DQDB) | Disbanded |
| IEEE 802.7 | Broadband LAN using Coaxial Cable | Disbanded |
| IEEE 802.8 | Fiber Optic TAG | Disbanded |
| IEEE 802.9 | Integrated Services LAN (ISLAN or isoEthernet) | Disbanded |
| IEEE 802.10 | Interoperable LAN Security | Disbanded |
| IEEE 802.11 | Wireless LAN (WLAN) & Mesh (WiFi certification) | Active |
| IEEE 802.12 | 100BaseVG | Disbanded |
| IEEE 802.13 | Unused | Reserved for Fast Ethernet development[b] |
| IEEE 802.14 | Cable modems | Disbanded |
| IEEE 802.15 | Wireless PAN | Active |
| IEEE 802.15.1 | Bluetooth certification | Active |
| IEEE 802.15.2 | IEEE 802.15 and IEEE 802.11 coexistence | Active |
| IEEE 802.15.3 | High-Rate wireless PAN (e.g., UWB, etc.) | Active |
| IEEE 802.15.4 | Low-Rate wireless PAN (e.g., ZigBee, WirelessHART, MiWi, etc.) | Active |
| IEEE 802.15.5 | Mesh networking for WPAN | Active |
| IEEE 802.15.6 | Body area network | Active |
| IEEE 802.15.7 | Visible light communications | Active |
| IEEE 802.16 | Broadband Wireless Access (WiMAX certification) | Active |
| IEEE 802.16.1 | Local Multipoint Distribution Service | Active |
| IEEE 802.16.2 | Coexistence wireless access | Active |
| IEEE 802.17 | Resilient packet ring | Hibernating |
| IEEE 802.18 | Radio Regulatory TAG | Active |
| IEEE 802.19 | Coexistence TAG | Active |
| IEEE 802.20 | Mobile Broadband Wireless Access | Hibernating |
| IEEE 802.21 | Media Independent Handoff | Active |
| IEEE 802.22 | Wireless Regional Area Network | Active |
| IEEE 802.23 | Emergency Services Working Group | Active |
| IEEE 802.24 | Smart Grid TAG | Active |
| IEEE 802.25 | Omni-Range Area Network | Not yet ratified |

[a]http://standards.ieee.org/about/get/
[b]https://en.wikipedia.org/wiki/IEEE_802#cite_note-3

**Table 13.2** Types of wireless networks by geographical range[a]

| Type | Range | Applications | Standards |
|------|-------|--------------|-----------|
| Personal area network (PAN) | Within reach of a person | Cable replacement for peripherals | Bluetooth, ZigBee, NFC |
| Local area network (LAN) | Within a building or campus | Wireless extension of wired network | IEEE 802.11 (WiFi) |
| Metropolitan area network (MAN) | Within a city | Wireless internetwork connectivity | IEEE 802.15 (WiMAX) |
| Wide area network (WAN) | Worldwide | Wireless network access | Cellular (UMTS, LTE, etc.) |

[a]http://grouper.ieee.org/groups/802/

Table 13.2 presents basic types of wireless networks based on their "geographic range".

*Personal Area Network*

A personal area network (PAN) is a computer network used for communication among computer and different information technological devices close to one person. Some examples of devices that are used in a PAN are personal computers, printers, fax machines, telephones, PDAs, scanners, and even video game consoles. A PAN may include wired and wireless devices. The reach of a PAN typically extends to 10 m. A wired PAN is usually constructed with USB and Firewire connections while technologies such as Bluetooth and infrared communication typically form a wireless PAN.

*Local Area Network*

A local area network (LAN) is a network that connects computers and devices in a limited geographical area such as home, school, computer laboratory, office building, or closely positioned group of buildings. Each computer or device on the network is a node. Current wired LANs are most likely to be based on Ethernet technology, although new standards like ITU-T G.hn also provide a way to create a wired LAN using existing home wires (fiber optic and coaxial cables, phone power lines).

*Metropolitan Area Network*

Metropolitan area network is a large computer network that can cover a city or any campus, not limited to any specific wire or wireless technology.

*Wide Area Network*

A wide area network (WAN) is a computer network that covers a large geographic area such as a city, country, or spans even intercontinental distances, using a communications channel that combines many types of media such as telephone lines, cables, and air waves. WAN often uses transmission facilities provided by common carriers, such as telephone companies. The WAN technologies generally function at the lower three layers of the OSI reference model: the physical layer, the data link layer, and the network layer.

The different scales of computer networks, and their various lay outs and operation modes, require specifically adjusted cyber security approaches and physical security measures.

## 13.7 Wireless Operating Modes

Wireless network adapters and access points may operate in several modes, allowing versatile functionality. There are four main communication modes in which the wireless network can be organized:

- Infrastructure mode
- Ad hoc mode
- Wireless distribution mode
- Monitor mode.

### 13.7.1 Infrastructure Network Mode

In the infrastructure network mode, there is at least one AP and one station, which together form a basic service set (BSS). The AP is most commonly connected to a wired network, which is called a distribution system (DS). Two or more wireless APs connected to the same wired network form an extended service set (ESS)—a single logical network segment.

### 13.7.2 Ad Hoc Network Mode

An ad hoc network, also known as an independent basic service set (IBSS), consists of at least two stations communicating without an AP. This mode is also called "peer to peer mode". In an ad hoc network, one of the participating stations adopts some of the responsibilities of an AP, such as:

- Beaconing
- Authentication of new clients joining the network.

In the ad hoc mode, the station fully trusts the AP without any verification.

### 13.7.3  Wireless Distribution Mode

A wireless distribution mode is similar to a standard distribution system, but it operates wirelessly and APs communicate with one another. Wireless distribution mode has two connectivity modes:

- Wireless Bridging: only allows wireless distribution mode APs to communicate with each other
- Wireless Repeating: allows both stations and APs to communicate with each other.

### 13.7.4  Monitor Mode

Monitor mode allows a wireless card to capture, or monitor, the packets which are received without any filtering. The tools for wireless security testing (*e.g., Airodump-ng, Aireplay-ng*) require the wireless adapter to be reconfigured in a monitor mode in order to perform the attack (monitor wireless traffic, create and inject custom packets, etc.).

## 13.8   Cyber Security Assessment Methodologies

As any sophisticated procedure, cyber security assessment has a variety of methodologies to specify testing procedures and reporting standards. The most popular of them are the following:

1. OSSTMM—Open Source Security Testing Methodology Manual
2. ISSAF—Information Systems Security Assessment Framework
3. NIST800-115—National Institute of Standard and Technology methodology.

*ISSAF*

The Information Systems Security Assessment Framework (ISSAF) is a peer reviewed structured framework designed by the Open Information Systems Security Group (OISSG). The methodology, defined by ISSAF, covers all aspects related to security assessments: from a high-level perspective (e.g., business impact and organizational models) to practical techniques (e.g., security testing of passwords, systems, network, etc.). The framework is divided into four main phases structured in four working packages: planning, assessment, treatment, and accreditation.[11]

---

[11]http://www.oissg.org/issaf.

*OSSTMM*

The Open Source Security Testing Methodology Manual (OSSTMM) is an open standard methodology for security tests.[12] It was developed in 2000 by Pete Herzog as a cyber security assessment framework and has rapidly become a comprehensive methodology to assure security at operational level. It encompasses tests for virtually every publicly known security aspect: from personnel qualification to physical security, from control of communication to electronic systems safety. As every standard methodology, it is designed to be consistent and repeatable. Moreover, it is openly available and thus allows its free dissemination and free use.

OSSTMM includes the following key sections:

- Operational Security Metrics
- Trust Analysis
- Work Flow
- Human Security Testing
- Physical Security Testing
- Wireless Security Testing
- Telecommunications Security Testing
- Data Networks Security Testing
- Compliance Regulations
- Reporting with the STAR (Security Test Audit Report).

*NIST SP800-115*

The National Institute of Standard and Technology (NIST) has designed its own security assessment methodology in the Special Publication 800-115 [6]. It provided guidelines for organizations on planning, conducting, and evaluating information security testing. The overall goal of NIST SP800-115 is to propose an overview of the main key elements of technical security assessments. It also provides practical recommendations and technical information related to penetration tests.

NIST SP800-115 focuses on singular tests and final reports. This methodology has a potential to compare the outcomes of various security performed from different "Testing Viewpoints" (i.e., different information provided by the auditors or different physical location of the cyber security testing experts). The methodology NIST800-115 defines four testing types:

- External Security Testing: type of testing that is conducted from outside the security perimeter of the tested facility. The goal of this security test is to reproduce the view of an external attacker and to draw the attention to vulnerabilities already visible from outside the company premises (e.g., from the Internet). External testing always begins with discovery techniques aiming at examining organization public presence.
- Internal Security Testing: the type of testing that is performed within an organization's perimeter (e.g., internal network). The purpose of the internal security

---

[12]http://www.isecom.org/research/osstmm.html.

testing is to impersonate a trusted insider or also an attacker that has penetrated external defenses. This test allows to have a certain level of access to the network and to internal information, in order to assess internal security mechanisms.

- Overt Security Testing: also known as white hat testing, defines an internal or external testing in which the experts have a full knowledge of organization's systems and processes. Company staff is also aware of the test and it usually tries to limit the testing impact. This kind of test is often used as company training.
- Covert Security Testing: also known as black hat testing, uses an adversarial approach by not providing any knowledge about the organization. In this scenario, company IT staff is usually also not aware of the test and its response is used for testing the technical and organizational security controls within the company.

According to the practical experience of major security companies and OWASP research,[13] ISSAF is considered as a very comprehensive reference source,[14] OSSTMM is the best standardized security testing framework[15] and NIST SP800-115 is the best suited for singular testing and structured reports.

## 13.9   Security Testing Practical Applications

Security testing practices and investigation techniques are based on developed standards and methodologies. However, each case requires individual approach and internal rules and procedures of any organization can influence the way in which the security testing is conducted.

To guarantee the best effect and to protect both parties of the agreement, a security expert team and an organization, any security testing should follow the preliminary agreed rules and procedures. Considering that each organization has its specific structure and challenges, there might be modifications in relation to the scale, tested location, types of equipment, operating system, findings before and during testing. All of them should be discussed in the pre-engagement stage while developing an agreement.

### 13.9.1   Preparatory Stage

Before initiating any practical step, a security assessment and testing require detailed planning. All potential challenges need to be discussed and agreed in

---

[13]https://www.owasp.org/index.php/Penetration_testing_methodologies.

[14]https://sourceforge.net/projects/isstf/files/issaf%20document/issaf0.1/.

[15]http://www.isecom.org/research/osstmm.html.

advance, as during the implementation stage the target system may reveal sensitive or even classified information. Even more, the system itself can also be damaged during exploitation or denial of service. Legal agreement and predefined procedure should be documented before the test begins.

### 13.9.1.1  Procedures, Documentation, and Limitations

Pre-engagement interactions include discussions and development of a documentation package, which should cover the following:

*Goals*

The "goal" should clearly state the final outcomes of the assessment. Among many goals there might be the following: test the compliance requirements, implemented controls, protection mechanisms, etc.

*Scope*

The "scope" documents which systems, networks, applications, processes, etc., are to be assessed and audited. This information will be important especially in the hosted environments, where the infrastructure may not be wholly owned by the client. The infrastructure components should be noted and ensured to be excluded from active penetration testing techniques.

*Testing Terms and Definitions*

To ensure overall understanding of the implemented procedure, it is important that all people involved have the same understanding of terminology used in the penetration The *Glossary* should be provided and discussed with the target audience of the final report, which may include a chief technology officer, a board of directors, internal IT teams, and an internal cyber security team.

*Communication channels*

Any communication with the client should be documented through officially approved channels. It is equally important for any unforeseen emergency, also for the official confirmation of each level implementation which might require introduction of changes in the initially developed plan, review, or assessment of findings.

*Terms of Reference*

The terms of reference should be developed to detail each step of the penetration test and to serve as a guidance. This document includes specification on when and how the security test is to be carried, what systems are permitted to be tested, and what actions should be performed with an exploited target to confirm the vulnerability. They will also include the approved time frames.

Depending on the type of security tests, the documentation of the facility incident response capabilities and capacities should be discussed and documented prior to the test.

## *Code of Conduct*

As the engagement of external experts might entail potential disclosure of sensitive information or any conflict of interest, a code of conduct for experts should outline responsibilities of experts to safeguard internal sensitive information.

As soon as the pre-engagement documentation is prepared, the hardware and software should be configured and deployed for testing.

### 13.9.1.2   Deployed Hardware

Hardware is the necessary perquisite to detect frequencies, identify protocols, connect to the network, and implement the attack.

The following hardware is relevant to wireless security testing:

- Computer system, laptop, smartphone, or a tablet
- Spectrum analyzers
- GPS Receivers
- Wireless antenna
- Rogue devices.

### Computers

A multi-core processor or multiprocessor computer is required for successful and fast password guessing (e.g., brute force attack, dictionary attack, etc.). The traffic capture is less resource intensive, and requires from the computer only its capability to store the captured data. However, the more intense the data transaction is in the network, the faster the computer will have to write the data to the hard drive. To conduct more sophisticated wireless attacks, the attack computer system multiple wireless adapters are required.

### Laptops

Laptop is a portable lightweight computer that requires only security testing operating system (e.g., Kali, Backbox, Black Arch, etc.) and a wireless antenna to prepare it to use as an attack platform.

### Smartphones and tablets

The Android devices (smartphones and tablets) can be used as penetration testing tools with custom penetration testing mobile operating systems (e.g., Kali NetHunter, PwnPhone, etc.) or downloaded network discovery applications.

### Spectrum Analyzers

Spectrum analyzers are high-end devices that monitor the strength of various frequencies in the RF spectrum. They are expensive and complicated to use, though they need to be operated by qualified personnel.

## GPS Receivers

GPS is a global navigation satellite system that provides geolocation and time information to a GPS receiver in all weather conditions, anywhere on or near the Earth where there is an unobstructed line of sight to four or more GPS satellites.[16] The GPS system operates independently of any telephonic or internet reception, though these technologies can enhance the usefulness of the GPS information. The GPS system provides critical positioning capabilities to military, civil, and commercial users around the world, and can be used by the security testing experts for mapping wireless coverage and equipment deployment. It is important to note that the GPS system does not work indoor, which makes its use irrelevant for surveys inside facilities.

GPS systems are vulnerable to "GPS spoofing", which may falsify directions in GPS-dependent systems (e.g., unmanned aerial vehicles, vessels, etc.).

## Wireless antennas

The wireless antennas and signal amplifiers allow to extend the range of wireless devices, permitting the testing team to simulate a cyberattack from the outside of facility (in some cases—miles away).

Below are the types of antennas, used by security experts and attackers[17]:

*Manufactured*:

*Dipole*—an omnidirectional antenna that radiates radio wave power uniformly in all directions in one plane. Its transmission power decreases with elevation angle above or below the plane and drops to zero on the antenna's axis.

*Parabolic*—a directional parabolic antenna that uses a parabolic reflector, a curved surface with the cross-sectional shape of a parabola, to direct the radio waves.

Yagi—a very directional antenna with an effective range over 3 km. A Yagi–Uda antenna, commonly known as a Yagi antenna, is a directional antenna consisting of multiple parallel elements in a line, usually half-wave dipoles made of metal rods.

*Improvised*:

*Pringles, cantenna* (can + antenna)—The improvised (Hacker Style, DIY) Yagi, a homemade directional waveguide antenna, made out of an open-ended metal can.[18]

*WindSurfer*—an improvised Parabolic antenna.[19]

---

[16]http://www.gps.gov/multimedia/tutorials/trilateration/.

[17]http://haifux.org/lectures/295/Haifux_wireless_hacking.pdf.

[18]http://www.lifehack.org/articles/lifehack/extend-your-wifi-range-using-diy-antenna-from-pringles-can.html.

[19]http://freeantennas.com/projects/template2/index.html.

**Rogue devices**

If the attackers have physical access to LAN cables, they can plant unauthorized devices and hardware backdoors (e.g., routers, antennas, repeaters, microcomputers, etc.), giving them the strategic advantage and direct access to the internal network.

## 13.9.2   Scanning and Enumeration Techniques

Scanning and enumeration are an essential part of the security testing that reveals exposed ports, processes and systems. In the wireless security testing both physical and software enumeration should be conducted.

*Hardware scanning and enumeration.*

To detect radio frequencies and thus approximate types of wireless networks active in the proximity, spectrum analyzers are used.

*Software scanning and enumeration.*

As soon as wireless frequency is detected, hardware configured and target is verified, software scanning is used to further enumerate AP types, devices connected to the wireless network, encryption used and possibly open ports and services, running on this AP devices.

## 13.9.3   Passive Traffic Capture and Identification

The passive listening to the wireless traffic without being associated with any actual network is implemented by a so-called monitor mode. The wireless adapters, that support the monitor mode, should be able to hear and capture any wireless traffic in the surroundings and not only within a specified network.

Depending on the type of traffic, a specific tool is used out of variety of tools available for the analysis (e.g., Wireshark, dsniff, driftnet, urlsnarf, msgsnarg, NetworkMiner, Chaosreader, foremost, xplico, etc.).

## 13.9.4   Simulated Attacks

The attacks on wireless network are the attempts to gain partial or full access to the network (e.g., authenticate with the AP without the knowledge of the access key or password). There are three main methods of authentication that are used in today's wireless LANs:

- open authentication

- shared authentication
- EAP (Extensible Authentication Protocol) authentication.

### 13.9.4.1  Attacks on Authentication

The *open authentication* method is the simplest among the three methods, as it requires only the end device is able to detect the service set identifier (SSID). When the SSID is known, the device can associate with the network. The drawback of open authentication is that the SSID is broadcast and can be identified.

The *shared authentication* method is commonly used in individual and small business wireless LAN implementations. This method uses a key that is pre-shared with both sides of the connection. The device is able to connect only upon verification of the key's identity.

The third method uses the Extensible Authentication Protocol (EAP) and is the most common method used in facility networks. The EAP method utilizes an authentication server that is queried for authentication using a variety of credential options.

Below are the most often implemented attacks on WiFi that bypass or exploit vulnerabilities in authentication and association.

Ramachandran [7], world renowned security researcher and the founder of Security Tube, enumerates cyberattack methods on WiFi as follows [7]:

- *ARP Replay*—The classic ARP request replay attack generates new initialization vectors (IVs). The attack software captures an ARP packet then retransmits it back to the access point. This causes the access point to repeat the ARP packet with a new IV. This process is repeated until enough IVs are collected to decrypt the WEP key.
- *Caffe-Latte*—This passive attack captures broadcast ARP requests.
- Hirte (Fragmentation Attack)—The attack software captures an ARP request or IP packet from the client. Once at least one is received, a small amount of data is extracted and then used to forge an ARP request packet targeted to the client. The attack is especially effective against ad hoc networks.
- ChopChop/KoRek—This attack allows to decrypt a WEP data packet without knowing the key (including networks with dynamic WEP). This attack requires at least one WEP data packet.
- FMS Attack—the Fluhrer, Mantin, and Shamir attack is a stream cipher attack on the widely used RC4 stream cipher. The attacker can recover the key in an RC4 encrypted stream from a large number of captured messages.
- PTW Attack—the Pyshkin, Tews, and Weinmann attack. The main advantage of the PTW approach is that it requires very few data packets to decrypt the WEP key.

### 13.9.4.2  Attacks on Wireless Cryptography

The encryption has always served as the best solutions to provide the layer of confidentiality in computer networks. The most widely used encryption protocols for wireless data transmission are WEP and WPA/WPA2.

**Attacks on WEP encryption**

The goal of an attacker is to discover the WEP shared secret key. For this purpose, a large number (usually 600,000) of frames need to be captured and processed. These frames use the same key. Considering a sufficient number of mathematically weak frames (distinguished from the large number of captured frames), the attacker performs systematic computation that exposes the bytes of the secret key. This computation may take a few seconds to minutes.

**Attacks on WPA encryption**

These attacks follow the following steps:

1. A De-Authentication signal sent to AP
2. AP re-authenticates with the Client
3. The attacker captures the Handshake
4. Attacker performs a dictionary or brute force attack on the captured Handshake.

In 2009, a Beck–Tews attack allowed to decrypt a WPA encrypted packet without knowing the key (Base on the ChopChop Attack), which confirmed the WPA encryption algorithm to be vulnerable.

It is considered, that there is no optimal way to crack the WPA2-AES encryption, making it the safest to use in corporate networks today. It is still susceptible to brute force password guessing, however this attack may take hours to years to successfully discover the key.

### 13.9.4.3  Attacks on Captive Portals

The additional web authentication was introduced to enhance security of wireless devices.

A captive portal is a web page, presented by a Layer 3 or Layer 2 of OSI model. It is shown to users before they gain broader access to the network services. It often presents a "landing" ("log-in") page that requires additional login and password credentials. The landing portal intercepts observed packets until the user is authorized to launch network sessions.

To bypass the captive portals and gain access to the network and/or Internet, in some implementations (F5, pfSense, Cisco WLC, etc.) it may be possible to forward traffic to a specific open port on the AP router. Spoofing MAC address of the currently or recently connected devices allows the attacker to use the existing login session.

#### 13.9.4.4 Attacks on Mesh Networks

Vulnerabilities in wireless mesh networks (WMN) [8] can be exploited by potential attackers to degrade or disrupt the network services. Limited security measures between the nodes, and the absence of centralized controls, make the protocols vulnerable at the link, network, and transport layers of the OSI model, as well as the application layer protocols becomes susceptible to malware propagation.

The attacks on physical and link layers of OSI model are inherently different.

The physical layer is responsible for frequency selection, carrier frequency generation, signal detection, modulation, and data encryption. As with any radio-based medium, a jamming attack in WMNs can be performed without sophisticated equipment and software. A jamming source may even be powerful enough to potentially disrupt communication in the entire network by strategically jamming sources, critical for HCWN.

The link layer of a WMN is vulnerable to passive eavesdropping, jamming, MAC address spoofing, replay, collisions in allocation and precomputation, etc.

The multi-hop wireless networks are prone to internal eavesdropping by the intermediate hops, wherein a malicious node captures all the data that it forwards without the knowledge of any other nodes in the network. Passive eavesdropping does not affect the network functionality directly, however, it leads to the compromise in confidentiality and integrity.

In the link layer jamming attack the attacker may transmit regular MAC frame headers (without payload) on the transmission channel which matches the MAC protocol being used in the victim network. Consequently, the legitimate nodes always find the channel overloaded and/or unreachable and pause for a random period of time before probing the channel again. This leads to the denial of service for the legitimate nodes. In addition to the MAC layer, jamming can also be used to exploit the network and transport layer protocols.

The intentional collision of frames occurs when two nodes attempt to transmit on the same frequency simultaneously. When frames collide, they are discarded and need to be retransmitted. An adversary may strategically cause collisions in specific packets such as acknowledgment control messages, causing denial of service.

#### 13.9.4.5 Attacks Against Common Security Appliances

There are popular security measures and recommendations that can be bypassed or easily breached by attackers and security testing teams. The following attack scenarios may be used and should be acknowledged for further security research.

*Detecting Not-beaconing AP*

Not-beaconing (Not-broadcasting) APs are not usually detected by standard devices in the standard operation modes. While in a monitoring mode, the adapter can detect the communication between the "stealth" AP and the device already connected to it.

The hidden wireless AP may also leak SSID Name. Depending on the config-uration, many wireless devices do not automatically connect to hidden networks. If the device has automatic connection enabled, it attempts to automatically probe the surroundings for the hidden wireless network revealing the SSID (that might be captured by putting the wireless adapter in monitor mode).

*Fake authentication*

The fake authentication attack on the WEP protocol allows an attacker to join a protected network, even without the secret key.

The first method "Open System authentication": a message is sent to an AP, requesting to join the network using Open System Authentication. The AP sends the reply, if the Open System authentication is allowed. The handshake does not require the encryption key, thus, WEP protected network can be joined.

The second method, Shared Key Authentication, uses the secret key and a challenge–response authentication mechanism, which was designed to be a secure alternative to Open System authentication. However, an attacker who is able to capture a full Shared Key Authentication handshake sequence can join the network itself.

*Bypassing Whitelisting*

APs with the enabled whitelisting access control, permit only devices (stations) with known MAC addresses. Either the computer system, connected to the AP, has to be compromised, or the MAC address of the attacking device has to be spoofed with a legitimate already whitelisted MAC address.

*Bypassing IDS*

Intrusion detection systems are a vital part of modern cybersecurity software measures. Impersonation of the legitimate internal computer (changing device, OS and software signatures to internal legitimate users, spoofing MAC address, etc.) may allow the attacker not to be detected by the IDS.

Encryption (SSH, SSL, etc.) and Covert (DNS response, ICMP reply, etc.) channels may be used to infect the target network with malware and issue com-mands covertly.

All real time IDS system can suffer from issuing false alarms, especially those that use the anomaly based approach. The monitoring of IDS alerts is also a 24/7 activity and relies on human intervention, as attackers are not confined with working hours. Moreover, the attackers strategically plan their attacks when the response team is unavailable. To illustrate complacency amongst staff, Bruce Schneier, in Secrets and Lies [9], tells the story of a 22 h eBay outage in 1999 "when the IDS system set off alarms constantly, but everyone was too busy to respond."

### 13.9.5  Post-Exploitation

Privilege escalation allows the attacker to gain higher access privileges and access data previously inaccessible. In highly critical networks even the lowest privilege level can compromise the mission.

Information gathering after the exploitation of the wireless router can enumerate the entire facility network, and allows the attacker to gain further unauthorized access.

After gaining access to one of the devices, the attacker can install backdoor malware to connect gain increased control over the system and simplify repetitive access in the future.

### 13.9.6  Reporting

Upon the finalization of the technical part of the security test a detailed report should be developed and submitted to the internal security department and senior management. The assessment report should contain the executive summary, the narrative part and annex with technical details.

*Executive Summary*

The executive summary outlines the overall findings of the assessment and provides a high-level overview of the vulnerabilities discovered. The wording should be adapted to nontechnical target audience and include illustration materials for easier visibility.

The developed report should emphasize the risk level and indicate that the identified vulnerabilities provide due input in the risk management process. As cyber security experts are not the top level business managers and are not necessarily acquainted with the reputational or any other consequences of successful attacks, their task is limited to informing and urging the senior management for action.

*Narrative Report*

The narrative report introduction should outline the parameters of the security testing, the findings, and remediation. It should cover the project objectives and the expected outcome of the assessment, scope and schedule, targets, challenges or limitations, and proposed solutions. Special attention should be paid to the detailed coverage of limitations. For example, limitations of project-focused tests, limitation in the security testing methods, performance or technical issues that the tester come across during the course of assessment, etc. The enumeration of findings should outline the vulnerabilities that were discovered during testing. Each finding should be clear and concise and give the reader of the report a full understanding of the challenge. A remediation summary should propose the action plan for fixing the vulnerabilities identified during testing.

*Annex with technical details*

This section includes detailed technical information about the vulnerabilities found and the actions needed to resolve them. It is aimed for the technical staff and should include all the necessary information, screenshot and command lines, affected items, severity rating, with vector notation if using CVSS, etc., for the technical teams to understand the issue and resolve it.

The confirmed vulnerability is reported to the software development company or to the company department for patch development and security updates. The duly developed security update should be assigned for prioritized installation during the next scheduled maintenance.

## 13.10   Vulnerability Management

The vulnerability management is the cyclical practice of identifying, classifying, remediating, and mitigating vulnerabilities, especially in software and firmware. Vulnerability management is integral to computer security and network security [10].

The wireless cybersecurity, being part of physical, computer and network security, required constant vigilance, regular risk management, incident response policies and planned tactics for change management.

### 13.10.1   Incident Response

The incident response is an organized approach to addressing and managing the aftermath of a security breach or attack (also known as an incident). The goal is to handle the situation in a way that limits damage and reduces recovery time and costs. An incident response plan includes a policy that defines, in specific terms, what constitutes an incident and provides a step-by-step process that should be followed when an incident occurs.

The immediate incident response is provided when:

- the attempt to breach is detected
- the breach is detected
- the long-term presence is detected
- the data leak happened
- the damage to system or equipment is detected.

Beyond software and equipment assessment, the external security teams may test the effectiveness of the internal incident response capabilities of the target organization, to audit its existing internal "controls" and "capabilities". To conduct this assessment, the testing team may employ social engineering (spearphishing emails, phone calls, physical infiltration, etc.).

### 13.10.2 Operational Security

In a more general sense, operational security is the process of protecting individual pieces of data that could be grouped together to give the bigger picture (called aggregation). OPSEC is the protection of critical information deemed mission essential from military commanders, senior leaders, management or other decision-making bodies. In civilian organization, in addition to generally established security procedures, project-specific operational security measures should be agreed and adopted.

### 13.10.3 Vulnerability Classification

According to Common Vulnerability Scoring System (CVSS),[20] vulnerabilities are design flaws or misconfigurations that make the network (or a host on the network) susceptible to malicious attacks from local or remote users. Vulnerabilities can exist in several areas of both the corporate and industrial networks, such as in firewalls, FTP servers, Web servers or operating system. Depending on the level of the security risk, the successful exploitation of a vulnerability can vary from the disclosure of information about the host to a complete compromise of the host. The severity levels for vulnerabilities are level 1 (minimal), level 2 (medium), level 3 (serious), level 4 (critical), and level 5 (urgent).

Potential vulnerabilities include vulnerabilities that cannot be fully verified. In these cases, at least one necessary condition for the vulnerability is detected. It is recommended to always investigate these vulnerabilities further. After the vulnerabilities are confirmed and classified, the contract agreement should specify if the proof-of-concept exploit be created to illustrate the danger of the vulnerability.

## References

1. US University of Peace. (2015). *Special report*
2. Morteza M. Zanjireh, A. S. (2013). *ANCH: A New Clustering Algorithm for Wireless Sensor Networks*. Glasgow Caledonian University
3. Donahue, G. A. (2007). *Network Warrior*. O'Reilly

---

[20]https://www.first.org/cvss/v2/guide.

4. NIST. (2004). *Standards of Security Categorization of Federal Information and Information Systems.* National Institute of Standards and Technology
5. IEEE. (2004). *Overview and Guide to the IEEE 802 LMSC.* IEEE
6. NIST. (2008). *Technical Guide to the Information Security Testing and Assessment.* National Institute of Standards and Technology Special Publication
7. V. Ramachandran, C. B. (2016). *Kali Linux Wireless Penetration Testing.* PACKT Publishing. Open Source
8. Sen, J. (2013). *Security and Privacy Issues in Wireless Mesh Networks: A Survey.* Innovation Labs, Cornell University
9. Schneier, B. (2000). *Secrets & Lies.* Wiley
10. Foreman, P. (2010). *Vulnerability Management.* Taylor & Francis Group

# Chapter 14
# Future Attack Patterns

As the wireless technologies are expending to every aspect of our lives, the advantages and the threats will only grow in the future.

## 14.1 Cyberattacks

It is assumed by default, that the future goals of the low-level cybercriminals will be the same, yet their arsenal and capabilities for uneducated non-professional cyberattacks will expand. Due protection should be enforced against the following cyber thefts:

- *Information theft*—with exponentially expanding data and human dependency on information threats to information misuse become more dangerous.
- *Financial theft*—with decreasing cash use and increasing popularization of electronic transactions, economic cybercrime will create new ways for criminal financial gain.
- *Identify theft*—with raise in biometrics and electronic presence, identity theft will be even easier to implement.

## 14.2 Hybrid Attacks

Hybrid attacks are the attacks that incorporate several inherently different types of attacks, combined with cyberattacks. They may be performed in the following environment.

© The Author(s) 2017
M. Martellini et al., *Information Security of Highly Critical Wireless Networks*,
SpringerBriefs in Computer Science, DOI 10.1007/978-3-319-52905-9_14

## 14.2.1   Against Facilities

*Water supply—threat to human health*

In October 2006, the attackers gained access to computer systems at a Harrisburg water treatment plant in the USA. The ICS network was accessed after an employee's laptop computer was compromised via the Internet, and then used as an entry point to install a malware that was capable of affecting the plant's water treatment operations.

*Power supply—threat to human well-being*

In December 2015, the Denial of Service in a power plant and multiple substations in Ukraine triggered a power outage. In February 2016, it was acknowledged that BlackEnergy3 malware was used for the cyberattack.[1]

*Heat supply—threat to human lives in the North*

In November 2016, the Distributed Denial of Service attack led to the disruption of the heating systems for at least two housing blocks in the city of Lappeenranta, Finland, literally leaving their residents in subzero weather. In an attempt to fight back the cyberattacks, which lived for a short time, the automated systems rebooted—and unfortunately got stuck in an endless loop, which restarted repeatedly and eventually shut down heating systems for more than a week.

Attacks on the civilian infrastructure may become more critical, undermining human well-being and security.

## 14.2.2   Against Consumer Products

Smart watch may be potentially used to launch malicious scripts while connected to wireless networks, initiating a cyberattack and/or malware propagation.

Embedded devices (Internet of Things) and Smart Homes are already targeted and have been recently (21 October 2016) used to generate the biggest DDoS attack in history as for present.[2] In the future, spying and unauthorized use of embedded devices will develop into more sophisticated attacks.

---

[1]http://www.ibtimes.com/us-confirms-blackenergy-malware-used-ukrainian-power-plant-hack-2263008.

[2]http://thehackernews.com/2016/10/iot-dyn-ddos-attack.html.

### 14.2.3  Against AWS

Autonomous Weapons Systems will revolutionize wars and law enforcement activities. Their key advantages are very obvious: the AWSs enhance the safety of human operators, work in 24/7, have better military capabilities (speed, resilience, accuracy, reaction time, flight, etc.), but at the current level of technological progress they do not guarantee the fidelity of performance.

Wireless cyberattacks may be conducted to override the AWS controls and issue false commands.

### 14.2.4  Against Unmanned Vehicles

The self-driving cars will make our life easier and provide more freedom in travel. However, they are already tested for cybersecurity vulnerabilities,[3] as the laptops are becoming more popular among the criminals stealing vehicles.[4]

The self-driving military patrol vehicles are controlled wirelessly or in an autonomous mode. Overriding controls and stealing a military patrol vehicle can give the terrorists resources to conduct sophisticated terrorist attacks without risking their own lives, however endangering lives of countless civilians. UAVs and UCAVs—overriding control or changing GPS coordinates of combat aerial vehicles can allow the attackers to disrupt a military operation, change the target of the UAV and/or cause the loss of civilian lives.

### 14.2.5  Against Satellites, Weaponization of the Outer-Space and Interplanetary Internet

It was in 1999, when the Telegraph published the following story, "A group of computer hackers suspected of seizing control of a British military communications satellite using a home computer, triggering a "frenetic" security alert has been traced to the south of England".[5]

The satellite security will have to be increasingly protected, and the potential weaponization of the outer-space might give birth to a new profession—a space cybersecurity agent.

---

[3]https://www.bloomberg.com/news/articles/2016-07-19/cybersecurity-is-biggest-risk-of-autono mous-cars-survey-finds.

[4]http://www.wsj.com/articles/thieves-go-high-tech-to-steal-cars-1467744606?mod=e2tw.

[5]http://www.sans.edu/research/security-laboratory/article/satellite-dos.

The interplanetary Internet (IPN, or InterPlaNet) is a theorized wireless communication and computer network in space, consisting of a set of network nodes. As the Internet is considered a very overloaded network of sub-networks with high traffic, the theorized interplanetary Internet is a store and forward network of multiple global area networks that is often disconnected, has a wireless architecture with delays ranging from tens of minutes to even hours, even when there is a connection. The data will be potentially transferred in bulk, making the retransmission equipment vulnerable to passive traffic capture, spoofing attacks, and denial of service, undermining confidentiality and integrity of interplanetary data transfer.

Future challenges may also create new opportunities for the security experts, and newly developed protocols may solve many current security problem that exist due to vulnerabilities in protocols of more than 20 years of age.

### 14.2.6   Against Medical Equipment

The attacked computerized medical surgical equipment may disrupt the surgical procedure and cause the loss of a patient's life. The implantable equipment (organs, pacemakers, etc) are already vulnerable to wireless attacks, as they use wireless control methods.

Medical and Military implants and implantable computers, potentially used for communication, reporting, targeting, advanced surveillance, may be used to capture traffic, capture communication feed, damage the operator (if the device has feedback or direct hardware access), etc.

# Chapter 15
# Assessing Cyberattacks Against Wireless Networks of the Next Global Internet of Things Revolution: Industry 4.0

## 15.1 Introduction

Historically, technology advances and increase in productivity led to revolutionary societal changes and industrial development. The first industrial revolution created machines to replace hand work and invented steam engine to decrease hard labor. The ambitious engineer thought catalyzed the second industrial revolution, electrification increased the working hours and assembly lines enhanced mass production in the beginning of the twentieth century. While its second part witnessed a real breakthrough in computer engineering, and industrial automation spread exponentially taking over the manual controls; cyberspace ensured global digital communication, mobile connection, and e-commerce. Electronics and internet technologies created the thirst industrial revolution. The beginning of the XXI century is operating with such realities as Internet of Things, Robotics, Virtual Reality, Cyber Warfare, and Industry 4.0.

The current development of emerging technologies allows to speak about the fourth wave of technological breakthrough: the rise of new digital industrial technology, defined by Henning Kagermannas et al.[1] as Industry 4.0, a transformation that is powered by nine foundational technology advances (see Fig. 15.1).

"The first three industrial revolutions came about as a result of mechanization, electricity and IT. Now, the introduction of the Internet of Things and Services into the manufacturing environment is ushering in a fourth industrial revolution. In the future, businesses will establish global networks that incorporate their machinery, warehousing systems and production facilities in the shape of Cyber-Physical Systems (CPS)", says the Final report of Industry 4.0 Working group[2] (2013). The report explains further that these systems will comprise smart machines, storage

---

[1]http://www.vdi-nachrichten.com/Technik-Gesellschaft/Industrie-40-Mit-Internet-Dinge-Weg-4-industriellen-Revolution.

[2]http://www.acatech.de/fileadmin/user_upload/Baumstruktur_nach_Website/Acatech/root/de/Material_fuer_Sonderseiten/Industrie_4.0/Final_report__Industrie_4.0_accessible.pdf.

© The Author(s) 2017
M. Martellini et al., *Information Security of Highly Critical Wireless Networks*,
SpringerBriefs in Computer Science, DOI 10.1007/978-3-319-52905-9_15

**Fig. 15.1** The nine pillars of Industry 4.0 (https://www. bcgperspectives.com/content/ articles/engineered_products_ project_business_industry_ 40_future_productivity_ growth_manufacturing_ industries/)

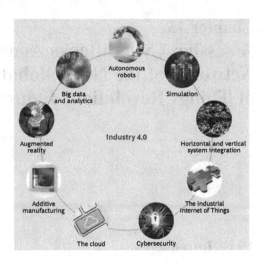

systems, and production facilities capable of autonomously exchanging information, triggering actions, and controlling each other independently.

The term "Industrie 4.0" has originated from the German project related to the development of high technology strategy for enhancing digitization of manufacturing.[3] Currently, Industry 4.0 is the modern trend of automation and data exchange in manufacturing, communication and control technologies. It includes cyber-physical systems, the Internet of Things and cloud computing. Industry 4.0 means overall smart architectural interconnectivity, or, a so called "smart factory." Within the modular structured facilities, cyber-physical systems monitor physical processes through the smart sensors, create a virtual duplicate of the physical world and make decentralized decisions. Cyber-physical systems communicate and cooperate with each other and with humans (operators and consumers) in real time. Internal and cross-organizational services are offered and used by participants of the supply chain, requiring increased interconnectivity between organizations.

In October 2012, the Working Group on Industry 4.0 presented a set of Industry 4.0 implementation recommendations to the German Federal Government [1]. The Industry 4.0 workgroup members are recognized by the German Federal Ministry of Education and Research as the founding fathers and driving force behind Industry 4.0. Currently Industry 4.0 concept is reviewed, planned and standardized by the following Workgroups:

- WG 1—The Smart Factory
- WG 2—The Real Environment
- WG 3—The Economic Environment
- WG 4—Human Beings and Work
- WG 5—The Technology Factor.

---

[3]https://www.bmbf.de/de/zukunftsprojekt-industrie-4-0-848.html.

The study conducted for the European Parliament [2], has already identified the challenges of the Fourth Industrial Revolution:

- Information Technology security issues are greatly aggravated by the inherent need to allow remote access to those previously disconnected production elements
- Reliability, stability and integrity are required for critical machine-to-machine communication, including very short and stable latency times
- Information Technology malfunctions mitigation, as those would cause expensive production outages and/or physical damage
- Industrial innovations protection (e.g., control and configuration files of the ICS, blueprints, unpublished innovative articles, etc.)
- Threat of redundancy of the corporate IT and Security departments
- Impact of business paradigm changes:

  - Sustainability and limits of export of new technologies are yet to be researched.
  - Vulnerability of the supply chain
  - Global competitiveness and European Union domestic manufacturing

- Controversial impact of social changes:

  - Lack of skilled and educated human resources to create the solid ground for fourth Industrial Revolution
  - Unemployment, caused by automatic processes and IT-controlled processes, especially for lower educated social groups.

All Industry 4.0 solutions, must ensure high resilience over smart, interconnected, and wireless networks. The system must support application in organizations, facilities, and industries of all scales but should also be able to efficiently respond to internal and external changes. Solutions must be built over the standards-compliant open architectures in which components can be modularly added, replaced or removed to meet the specific demands in the production process. Modular processes, for instance, permit facility engineers to assemble and re-provision their production equipment with components, without the need to change the entire system (e.g., ad hoc Wireless Routers and Access Points).

But as with all innovations, there are positives and negative aspects, as well as there are malicious actors, ready to exploit them. Cybercriminals and terrorists are actively preparing their resources to ensure that they get access to the Industry 4.0. Techniques have already proved to be efficient over the years (i.e., social engineering). They are even more powerful against the systems that are functional 24/7, as the highly connected and interconnected Smart Factory is. Only Smart Security and Intelligent Intrusion Detection Systems can mitigate the risk of cyberattacks on the production facilities, critical infrastructure, military installations, and private Smart Houses.

## 15.2    Selected Security Threats of the Industry 4.0

Many organizations still rely on management and production systems that are unconnected or closed. With the increased connectivity and use of popular standard communications protocols that come with Industry 4.0, the need to protect critical ICS and manufacturing lines from cybersecurity threats increases dramatically. As a result, secure, reliable communications as well as sophisticated identity and access management of machines and users are essential.

Industry 4.0 participants suffer many of the same cyber threats as other organizations do. They have to counter the same external and insider threats as all businesses, of all sizes, have to contend with in current sophisticated cybersecurity landscape. They do however have some threats while not unique to Industry 4.0, are an issue. The following are some of those threats that Industry 4.0 players need to focus on and mitigate.

## 15.3    Advanced Persistent Threats and Cyber-Espionage

The use of security threat mechanisms known as an Advanced Persistent Threat (APT) in the manufacturing and military sectors is well known. APT's have been used for many years as the way to perform cyberattacks against a specific target, over a long period of time, to persist inside the target network and extract sensitive data. We are now seeing the development of well-funded (e.g., state actors) cybercriminal groups who use APT's to conduct cyber-espionage [3]. These groups are known to be highly skilled, well-coordinated and mobile, and they prioritize proprietary information and intellectual property.

Example: the Black Vine group who focuses on industries such as aerospace and utilities, Sandworm group—on ICS malware (BlackEnergy). Many of these types of groups exist—for example, in a recent series of allegations, the US steel industry has accused the Chinese government of stealing intellectual property though a sustained hacking campaign, which is likely to affect Chinese imports. The groups like Black Vine often use APT type malware that exploits zero-day vulnerabilities to slowly but covertly extract, often over months, confidential, sensitive, and/or classified data.

Industry 4.0 is more vulnerable to cyber-espionage because of the smart and connected business processes that underlie it so we are likely to see this type of cyber threat increase.

## 15.4   Cyber-Terrorism

The definition of cyber-terrorism covers a multitude of impacts, from data exposure to physical damage. Most often it can be seen to be politically motivated. Current terrorist groups are known to be actively working on cyber-terrorist techniques and

ISIS has a dedicated social media forum where adepts exchange cybersecurity information on how to create a catastrophic effect on critical infrastructure components such as utilities. An analysis on the threat of cyber-terrorism by ISIS, "Risks of ISIS-Cyber-Terrorism" [4] claims that one of the key determinants was the fact that Industry 4.0 is Internet enabled, and thus vulnerable to incoming cyberattacks with virtual and physical impact.

## 15.5  Supply Chain and the Extended Eco-System

One of the key features of Industry 4.0 is the ability to interconnect across environments, which has the potential to make the supply chain more efficient. However, supply chain security and cybersecurity issues are well known and exploited to great extent by cybercriminals. Many of the biggest security breaches have been initiated through a supplier, often by spearphishing and stolen privileged credentials, resulting in mass data exposure. Industry 4.0 gives the cybercriminal more opportunity to exploit the supply chain, reaching the Smart factory internal network through its dependent actors.

The communication infrastructure (including wireless and sensor networks) forms the backbone of all Smart factory concepts. Ensuring its secure and reliable operation is therefore the prerequisite for successful realization of the Industry 4.0 vision. Considerable efforts will still be needed before some of the required systems are defined, designed and deployed, however many of these activities are already ongoing, and in some cases existing solutions from other areas can be adopted and applied to industry applications, medical facilities, military installations, and critical infrastructure.

Only by utilizing modern cybersecurity counter measures (e.g., adaptive authentication and behavioral analysis), the flow of supply chain of the Industry 4.0 can be secured.

## 15.6  Challenges of the Internet of Things

The Internet of Things (IoT) is the internetworking of physical devices, vehicles (also referred to as "connected devices" and "smart devices"), buildings, factories, homes, and other items—embedded with electronics, software, sensors, actuators, and network connectivity that enable these objects to collect and exchange data.

The Internet of Things architecture creates a multitude of potential points of entry that can be potentially exploited by the malicious actors. The IoT has issues with security at a low level, which may be inherited by the newer versions of the vulnerable products (e.g., Microsoft Windows Atom Tables vulnerability affects all versions of Windows Operating Systems[4]), as the IoT is used to underpin Industry 4.0 processes.

The latest example, in November 2016, security researchers from Keen Lab were able to remotely infiltrate the Smart Car (Tesla, Model S) parking and driving controls through the owner's smartphone.[5] Should this interconnectivity be unprotected in industrial implementations of the Internet of Things, the cybersecurity breach may lead to the physical damage and loss of human lives.

The global security expert, Bruce Schneier, in his blog post on IoT and highly connected devices "When hacking could enable murder",[6] stated that we are in a situation whereby we have "[…] retrofit security in after the fact". Not having a security layer built into the IoT as a prerequisite, has exposed the whole system open to critical vulnerabilities, attracting cybercriminals. The reliance of Industry 4.0 means that manufacturing companies will inherit these vulnerabilities unless we take special precautions, such as more adaptive authentication measures and different levels of trust in machine-to-machine communication, to mitigate them.

## 15.7  Autonomous Weapon Systems and Robots

Manufacturers in many industries have long used robots to solve complex assignments, but robots are evolving for even greater utility. They are becoming more autonomous, flexible, and cooperative. Eventually, they will interact with one another and work safely side by side with humans and learn from them.

For example, Kuka, a European manufacturer of robotic equipment, offers autonomous robots that interact with one another. These robots are interconnected wirelessly so that they can work together and automatically adjust their actions to fit the next unfinished product in line. Similarly, industrial-robot supplier ABB is launching a two-armed robot (YuMi) that is specifically designed to assemble products (such as consumer electronics) alongside humans. One "arm" represents a single robot, and two arms are linked and programmed to cooperate in solving a single objective.

---

[4]https://isc.sans.edu/forums/diary/Windows+Atom+Bombing+Attack/21651/.
[5]http://thehackernews.com/2016/11/hacking-tesla-car.html.
[6]https://www.schneier.com/essays/archives/2016/01/when_hacking_could_e.html.

# References

1. Henning Kagermann, W. W. (2013). *Recommendation for implementating the strategic initiative INDUSTRIE 4.0.* German Ministry of Education and Research
2. Jan Smit, S. K. (2016). *Industry 4.0.* Directorate General for Internal Policies, European Parliament
3. Abaimov, S. (2015). *Advanced Persistent Threat: Stealth of Presence and Big Data Exfiltration.* Royal Holloway, University of London
4. Hilse, L. G. (2014). *Risks of ISIS-Cyber-Terrorism.* LARSHILSE

# Chapter 16
# Conclusion

Highly critical wireless networks (HCWN) are commercial wireless systems using TCP–IP protocols, and they are dominating our communications infrastructure. They are increasingly part of our industrial base, and will become more embedded in our daily lives as we take advantage of portable medical equipment, SCADA communications for critical utilities, and more efficient, connected household appliances and the Internet of Things. The conveniences provided by HCWN also come with new cybersecurity vulnerabilities. Protection against interception of messages can be provided through the use of encryption. However, other vulnerabilities may require new policies for protection against attacks against cyber vulnerabilities. Criminals must be prevented from monitoring electrical usage in real time to determine when a house is occupied or vacant. Medical devices must be manufactured with high security standards, and healthcare staff must be trained to follow effective practices to reduce cybersecurity vulnerabilities.

The wireless technologies, since their birth in 1890, have revolutionized our life style and provided freedom and liberty in remote data procession and management, both in civilian and military areas. After more than a century of evolution, wireless technology is adopted globally due to its advantages and now surrounds us on a daily basis.

HCWN are now part of every industry and work environments. They are commercial wireless systems and are regularly used by first responders and the military to support communications in remote areas. They are also used to control portable medical devices, connect remote locations with SCADA systems, and monitor household devices as they are part of the new smart electric grid for power distribution.

The new advantages are followed by the emerging cybersecurity challenges. Interdependencies and complexities of industrial and corporate wireless systems, designed to speed up the work efficiency, generate cyber security vulnerabilities by default as the access to wireless devices immediately grants the attackers access to internal networks. Consequently, in highly critical networks even the lowest privilege level can compromise the mission.

© The Author(s) 2017
M. Martellini et al., *Information Security of Highly Critical Wireless Networks*,
SpringerBriefs in Computer Science, DOI 10.1007/978-3-319-52905-9_16

The increasing sophistication of wireless devices, accidental and intentional misconfigurations of the equipment and exponentially growing number of vulnerabilities urge for research, monitoring, assessment, and testing of the wireless equipment and software. Though numerous organizations today develop wireless technology security and standards, the different scales of computer networks, and their various lay outs and operation modes, require specifically adjusted cyber security approaches and physical security measures.

Though basic security measures for wireless network have already been developed and successfully deployed, it is important to regularly upgrade them, monitor, and assess their implementation. Especially crucial it is for mission critical communications and military, medical, and CBRNe infrastructure.

Cyber security testing detects and classifies flaws in cybersecurity. It identifies risks in the system and/or network and classifies potential vulnerabilities to further help in developing software, hardware, or physical solutions. As the wireless systems have a more sophisticated design than the wired computer systems, their security testing is more complicated and includes acknowledgement of detectability and vulnerability of routers and adapters to further develop and deploy preventive measures against "eavesdropping," denial of service, security breaches, and unauthorized remote control of wireless devices.

Though security testing practices and investigation techniques are based on developed standards and methodologies, each case requires individual approach, and internal rules and procedures of any organization can influence the way in which the security testing is conducted. Before initiating any practical step, a security assessment and testing require detailed planning, as during the implementation stage the target system may reveal sensitive or even classified information, or the system itself can also be damaged during exploitation or denial of service. Legal agreement and predefined procedure should be documented before the test begins.

The preparatory stage develops the legal agreement and specific documentation such as Terms of Reference, Communication Channels, Codes of Conduct. The next step is configuration of hardware and software and their deployment for testing. The following hardware is relevant to wireless security testing: computer system, laptop, smartphone or a tablet, spectrum analyzers, GPS receivers, wireless antenna, rogue devices. Physical and software scanning and enumeration, that reveals exposed ports, processes, and systems, should be also conducted in wireless security test.

The attacks on wireless network are aimed to gain partial or full access to the network. There are three main methods of authentication that are used in today's wireless LANs: open authentication, shared authentication, EAP (Extensible Authentication Protocol) authentication. Among the different types of cyberattacks, there are attacks on captive portals, attacks on mesh networks, attacks against common security appliances: detecting not beaconing AP, fake authentication, bypassing whitelisting, IDS, intrusion detection systems.

After the finalization of the security testing, a detailed assessment report should be developed and submitted to the senior management. The confirmed vulnerability is reported to the software development company or to the company department, for patch development and security updates in vulnerability management.

The vulnerability management is integral to computer security and network security and includes identifying, classifying, remediating, and mitigating vulnerabilities. The incident response is an organized approach to addressing and managing the aftermath of a security breach or attack. The operational security is the protection of critical information considered mission or project essential.

As the wireless technologies are evolving, the advantages and the threats will only grow in the future. It is assumed by default, that the future goals of the low-level cybercriminals will be the same, yet their arsenal and capabilities for nonprofessional cyberattacks will expand. Due protection should be enforced against information, financial, identify cyber thefts. The sophisticated cyber attacks will increase vulnerability of all our life: facilities (water, power supply, etc.), consumer products, unmanned vehicles, satellites, medical equipment, IPN, and Internet of Things.

The future development will create new challenges, but will also provide a solid ground for new protocols and revolutionized wireless security solutions.

Printed in the United States
By Bookmasters